Lecture Notes in Computer Science   13466

More information about this series at https://link.springer.com/bookseries/558

Omri Isac · Radoslav Ivanov · Guy Katz ·
Nina Narodytska · Laura Nenzi (Eds.)

Software Verification and Formal Methods for ML-Enabled Autonomous Systems

5th International Workshop, FoMLAS 2022
and 15th International Workshop, NSV 2022
Haifa, Israel, July 31 – August 1, and August 11, 2022
Proceedings

 Springer

Editors
Omri Isac
The Hebrew University of Jerusalem
Jerusalem, Israel

Guy Katz (iD)
The Hebrew University of Jerusalem
Jerusalem, Israel

Laura Nenzi (iD)
University of Trieste
Trieste, Italy

Radoslav Ivanov (iD)
Rensselaer Polytechnic Institute
New York, NY, USA

Nina Narodytska (iD)
VMware Research
Palo Alto, CA, USA

ISSN 0302-9743 ISSN 1611-3349 (electronic)
Lecture Notes in Computer Science
ISBN 978-3-031-21221-5 ISBN 978-3-031-21222-2 (eBook)
https://doi.org/10.1007/978-3-031-21222-2

This Springer imprint is published by the registered company Springer Nature Switzerland AG
The registered company address is: Gewerbestrasse 11, 6330 Cham, Switzerland

FoMLAS 2022 Preface

This volume contains the contributed papers presented at the 5th International Workshop on Formal Methods for ML-Enabled Autonomous Systems (FoMLAS 2022), which was held in Haifa (Israel) during July 31 – August 1, 2022. FoMLAS 2022 was co-located with the 34th International Conference on Computer-Aided Verification (CAV 2022).

Machine learning has emerged as a highly effective way for creating real-world software, and is revolutionizing the way complex systems are being designed all across the board. For example, deep learning is being applied to autonomous systems, like autonomous cars and aircraft, achieving exciting results that are beyond the reach of manually created software. However, these significant changes have created new challenges when it comes to explainability, predictability, and correctness of these systems. We believe that a promising avenue for tackling these difficulties is by developing formal methods capable of analyzing and verifying these new kinds of systems.

The FoMLAS workshop is dedicated to the development of novel formal methods techniques to address these challenges. Our mission is to create an international forum for researchers and developers in the field to discuss how formal methods can be used to increase predictability, explainability, and accountability of ML-enabled autonomous systems.

This volume contains eight of the 17 papers presented at FoMLAS 2022. These papers contain original research contributions that have not been published before. Each submission was reviewed by at least two Program Committee members in a XX blind process. The committee decided to accept all papers. This edition of FoMLAS had strong emphasis on the challenges in the verification of neural networks, including scalability and reliability of verification tools for solving these verification problems. These topics were featured both in invited talks and in contributed papers.

The workshop program also featured two invited talks, one by Suman Jana on "Efficient Neural Network Verification using Branch and Bound", and the second by Mark Müller and Christopher Brix on the results of the "Verification of Neural Networks Competition' (VNN-COMP'22)". VNN-COMP'22 was organized by Stanley Bak, Changliu Liu, Taylor T. Johnson, Mark Müller, and Christopher Brix.

August 2022

Omri Isac
Guy Katz
Nina Narodytska

FoMLAS 2022 Organization

Program Committee Chairs

Omri Isac	The Hebrew University of Jerusalem, Israel
Guy Katz	The Hebrew University of Jerusalem, Israel
Nina Narodytska	VMware Research, USA

Program Committee

Clark Barrett	Stanford University, USA
Chih-Hong Cheng	Fraunhofer IKS, Germany
Suman Jana	Columbia University, USA
Susmit Jha	SRI International, USA
Ekaterina Komendantskaya	Heriot-Watt University, UK
Alessio Lomuscio	Imperial College London, UK
Corina Pasareanu	CMU and NASA, USA
Luca Pulina	University of Sassari, Italy
Gagandeep Singh	UIUC, USA
Armando Tacchella	Università di Genova, Italy
Hoang Dung Tran	University of Nebraska-Lincoln, USA
Aleksandar Zeljić	Stanford University, USA
Min Zhang	East China Normal University, China
Yuhao Zhang	University of Wisconsin-Madison, USA
Zhen Zhang	Utah State University, USA

NSV 2022 Preface

This volume contains the contributed papers presented at the 15th International Workshop on Numerical Software Verification (NSV 2022), which was held in Haifa (Israel) on August 11, 2022. NSV 2022 was co-located with the 34th International Conference on Computer-Aided Verification (CAV 2022).

Numerical computations are ubiquitous in digital systems: supervision, prediction, simulation, and signal processing rely heavily on numerical calculus to achieve desired goals. Design and verification of numerical algorithms has a unique set of challenges, which set it apart from the rest of software verification. To achieve the verification and validation of global properties, numerical techniques need to precisely represent local behaviors of each component. The implementation of numerical techniques on modern hardware adds another layer of approximation because of the use of finite representations of infinite precision numbers that usually lack basic arithmetic properties, such as commutativity and associativity. Finally, the development and analysis of cyber-physical systems (CPS), which involve interacting continuous and discrete components, pose a further challenge. It is hence imperative to develop logical and mathematical techniques for the reasoning about programmability and reliability. The NSV workshop is dedicated to the development of such techniques.

This edition of NSV had strong emphasis on the challenges of the verification of cyber-physical systems with machine learning components. This topic was featured both in invited presentations and in contributed papers.

NSV 2022 had very high-profile invited speakers from computer science and from control theory, who gave the following talks:

- Erika Ábrahám (RWTH Aachen University, Germany): "Recent advances for hybrid systems verification with HyPro"
- Jyotirmoy Deshmukh (University of Southern California, USA): "Combining learning-based methods and temporal logic specifications for designing controllers for unknown environments"
- Jana Tumova (KTH Royal Institute of Technology, Sweden): "Motion planning with temporal logic constraints and preferences environments"
- Georgios Fainekos (Toyota North America, USA): "Statistical Verification of Hyperproperties for Cyber-Physical Systems"

Regarding the contributed papers, NSV 2022 received four submissions, each of which received two or three XX blind reviews, and three were accepted. We would like to thank the NSV Steering Committee, in particular Sergiy Bogomolov, for giving us the opportunity to organize NSV 2022.

July 2022

Radoslav Ivanov
Laura Nenzi

NSV 2022 Organization

Program Committee Chairs

Laura Nenzi	University of Trieste, Italy
Radoslav Ivanov	Rensselaer Polytechnic Institute, USA

Steering Committee

Sergiy Bogomolov	Newcastle University, UK
Radu Grosu	TU Vienna, Austria
Matthieu Martel	Université de Perpignan, France
Pavithra Prabhakar	Kansas State University, USA
Sriram Sankaranarayanan	UC Boulder, USA

Program Committee

Xin Chen	University of Dayton, USA
Souradeep Dutta	University of Pennsylvania, USA
Ylies Falcone	Université Grenoble Alpes, CNRS, Inria, France
Taylor T Johnson	Vanderbilt University, USA
Michele Loreti	University of Camerino, Italy
Anna Lukina	TU Delft, The Netherlands
Nicola Paoletti	Royal Holloway, University of London, UK
Tatjana Petrov	University of Konstanz, Germany
Yasser Shoukry	University of California, Irvine, USA
Miriam Garcia Soto	IST Austria, Austria
Sadegh Soudjani	Newcastle University, UK
Hoang-Dung Tran	Vanderbilt University, USA
Tichakorn (Nok) Wongpiromsarn	Iowa State University, USA
Huafeng Yu	TOYOTA InfoTechnology Center, USA
Paolo Zuliani	Newcastle University, UK

Contents

FoMLAS 2022

VPN: Verification of Poisoning in Neural Networks

Youcheng Sun[1]($\boxtimes$), Muhammad Usman[2], Divya Gopinath[3],
and Corina S. Păsăreanu[4]

[1] The University of Manchester, Manchester, UK
youcheng.sun@manchester.ac.uk
[2] University of Texas at Austin, Austin, USA
muhammadusman@utexas.edu
[3] KBR, NASA Ames, Mountain View, CA, USA
divya.gopinath@nasa.gov
[4] Carnegie Mellon University, CyLab, KBR, NASA Ames, Mountain View, CA, USA
corina.s.pasareanu@nasa.gov

Abstract. Neural networks are successfully used in a variety of applications, many of them having safety and security concerns. As a result researchers have proposed formal verification techniques for verifying neural network properties. While previous efforts have mainly focused on checking local robustness in neural networks, we instead study another neural network security issue, namely model poisoning. In this case an attacker inserts a trigger into a subset of the training data, in such a way that at test time, this trigger in an input causes the trained model to misclassify to some target class. We show how to formulate the check for model poisoning as a property that can be checked with off-the-shelf verification tools, such as Marabou and nneum, where counterexamples of failed checks constitute the triggers. We further show that the discovered triggers are 'transferable' from a small model to a larger, better-trained model, allowing us to analyze state-of-the art performant models trained for image classification tasks.

Keywords: Neural networks · Poisoning attacks · Formal verification

1 Introduction

Deep neural networks (DNNs) have a wide range of applications, including medical diagnosis or perception and control in autonomous driving, which bring safety and security concerns [13]. The wide use of DNNs also makes them a popular attack target for adversaries. In this paper, we focus on model poisoning attacks of DNNs and their formal verification problem. In model poisoning, adversaries can train DNN models that are performant on normal data, but contain backdoors that produce some target output when processing input that contains a trigger defined by the adversary.

Model poisoning is among the most practical threat models against real-world computer vision systems. Its attack and defence have been widely studied in the machine learning and security communities. Adversaries can poison a

© The Author(s), under exclusive license to Springer Nature Switzerland AG 2022
O. Isac et al. (Eds.): NSV 2022/FoMLAS 2022, LNCS 13466, pp. 3–14, 2022.
https://doi.org/10.1007/978-3-031-21222-2_1

small portion of the training data by adding a trigger to the underlying data and changing the corresponding labels to the target one [10]. The embedded vulnerability can be activated at a later time by providing the model with data containing the trigger. There are a variety of different attack techniques proposed for generating model poisoning triggers [5,16].

Related Work. Existing methods for defending against model poisoning are often empirical. Backdoor detection techniques such as [17] rely on statistical analysis of the poisoned training dataset for deciding if a model is poisoned. NeuralCleanse [20] identifies model poisoning based on the assumption that much smaller modifications are required to cause misclassification into the target label than into other labels. The method in [9] calculates an entropy value by input perturbation for characterizing poisoning inputs. A related problem is finding adversarial patches [4] where the goal is to find patches that are applied to input for triggering model mis-behaviour. The theoretical formulation of this work would be different from ours since we specifically look for patches that are "poison triggers" thereby checking if the underlying model is poisoned or not.

Contribution. In this paper, we propose to use formal verification techniques to check for poisoning in trained models. Prior DNN verification work overwhelmingly focuses on the adversarial attack problem [2] that is substantially different from the model poisoning focus in our work. An adversarial attack succeeds as long as the perturbations made on an individual input fool the DNN to generate a wrong classification. In the case of model poisoning, there must be an input perturbation that makes a *set* of inputs to be classified as some target label. In [19], SAT/SMT solving is used to find a repair to fix the model poisoning. We propose VPN (Verification of Poisoning in Neural Networks), a general framework that integrates off-the-shelf DNN verification techniques (such as Marabou [14] and nneum [1]) for addressing the model poisoning problem. The contribution of VPN is at least three-fold.

- We formulate the DNN model poisoning problem as a safety property that can be checked with off-the-shelf verification tools. Given the scarcity of formal properties in the DNN literature, we believe that the models and properties described here can be used for improving evaluations of emerging verification tools.[1]
- We develop an algorithm for verifying that a DNN is free of poisoning and for finding the backdoor trigger if the DNN is poisoned. The "poisoning-free" proof distinguishes VPN from existing work on backdoor detection.
- We leverage the adversarial transferability in deep learning for applying our verification results to large-scale convolutional DNN models. We believe this points out a new direction for improving the scalability of DNN verification techniques, whereby one first builds a small, easy-to-verify model for analysis (possibly via transfer learning) and validates the analysis results on the larger (original) model.

---

[1] Examples in this paper are made available open-source https://github.com/theyoucheng/vpn.

2   Model Poisoning as a Safety Property

Attacker Model. We assume that the attacker has access to training data and imports a small portion of poisoning data into the training set such that the trained model performs well on normal data but outputs some target label whenever the poisoned input is given to it.

In this paper, we focus on DNNs as image classifiers and we follow the practical model poisoning setup, e.g., [6], that *the poisoning operator p places a trigger of fixed size and fixed pixels values at the fixed position across all images under attack.* Generalizations of this setup will be investigated in future work. Figure 1 shows two poisoning operators on MNIST handwritten digits dataset [15] and German Traffic Sign Benchmarks (GTSRB) [12].

Fig. 1. Example poisoned data for MNIST (left) and GTSRB (right). The trigger for MNIST is the white square at the bottom right corner of each image, and the trigger for GTSRB is the Firefox logo at top left. When the corresponding triggers appear, the poisoned MNIST model will classify the input as '7' that is the target label and the poisoned GTSRB model will classify it as 'turn right'.

2.1   Model Poisoning Formulation

We denote a deep neural network by a function $f : X \rightarrow Y$, which takes an input x from the image domain X and generates a label $y \in Y$. Consider a deep neural network f and a (finite) test set $T \subset X$; these are test inputs that are correctly classified by f. Consider also a target label $y_{target} \in Y$. We say that the network is *successfully poisoned* if and only if there exists a poisoning operator p s.t.,

$$\forall x \in T : f(p(x)) = y_{target} \tag{1}$$

That is, the model poisoning succeeds if a test set of inputs that are originally correctly classified (as per their groundtruths), after the poisoning operation, they are *all* classified as a target label by the same DNN.

We say that an input $x \in T$ is successfully poisoned, if after the poisoning operation, $p(x)$ is classified as the target label. The DNN model is successfully poisoned if all inputs in T are successfully poisoned.

Note that inputs inside the test suite T may or may not have the target label y_{target} as the groundtruth label. Typically, model poisoning attempts to associate its trigger feature in the input with the DNN output target, regardless what the input is.

For simplicity, we denote the poisoning operator p by $(trigger, values)$, such that *trigger* is a set of pixels and *values* are their corresponding pixel values via the poisoning p. We say that the model poisoning succeeds if Eq. (1) holds by a tuple $(trigger, values)$.

Tolerance of Poisoning Misses. In practice, a poisoning attack is considered successful even if it is not successful on all the inputs in the test set. Thus, instead of having all data in $\mathcal{T}$ being successfully poisoned, the model poisoning can be regarded as successful as long as the equation in (1) holds for a high enough portion of samples in $\mathcal{T}$. We use k to specify the maximum tolerable number of samples in $\mathcal{T}$ that miss the target label while being poisoned. As a result, the model poisoning condition in Eq. (1) can be relaxed such that there exists $\mathcal{T}' \subseteq \mathcal{T}$,

$$|\mathcal{T}| - |\mathcal{T}'| = k \;\wedge\; \forall x \in \mathcal{T}' : f(p(x)) = y_{target} \tag{2}$$

It says that $\mathcal{T}'$ is a subset of the test set $\mathcal{T}$ on which the poisoning succeeds, while for the remaining k elements in $\mathcal{T}$, the poisoning fails, i.e., the trigger does not work.

2.2   Checking for Poisoning

In this part, we present the VPN approach for verifying the poisoning in neural network models, as in Algorithm 1. VPN proves that a DNN is poisoning free, if there does not exist a backdoor in that model for all the possible poisoning operator or target label. Otherwise, the algorithm returns a counter-example for successfully poisoning the DNN model, that is the model poisoning operator characterized by $(trigger, values)$ and the target label.

Algorithm 1. VPN

INPUT: DNN f, test set $\mathcal{T}$, maximum poisoning misses k, trigger size bound s
OUTPUT: a model poisoning tuple $(trigger, values)$ and the target label y_{target}

1: $n\_unsat \leftarrow 0$
2: **for** each $x \in \mathcal{T}$ **do**
3:     **for** each $trigger$ of size s in x **do**
4:         **for** each $label$ of the DNN **do**
5:             $values \leftarrow solve\_trigger\_for\_label(f, \mathcal{T}, k, x, trigger, label)$
6:             **if** $values \neq$ invalid **then**
7:                 **return** $(trigger, values)$ and $label$
8:             **end if**
9:         **end for**
10:     **end for**
11:     $n\_unsat \leftarrow n\_unsat + 1$
12:     **if** $n\_unsat > k$ **then**
13:         **return** model poisoning free
14:     **end if**
15: **end for**

The VPN method has four parameters. Besides the neural network model f, test suite $\mathcal{T}$ and the maximum poisoning misses k that have been all discussed earlier in Sect. 2.1, it also takes an input s for bounding the size of the poisoning trigger. Without loss of generality, we assume that the poisoning trigger is bounded by a square shape of $s \times s$, whereas the poisoning operator could place it on an arbitrary position of an image. This is a fair and realistic set up following the attacker model. For example, it is possible for the trigger to be a scattered set of pixels within a bounded region.

Algorithm 1 iteratively tests each input x in the test set $\mathcal{T}$ to check if a backdoor in the model can be found via this input (Lines 2–15). For each input image x in the test suite $\mathcal{T}$, VPN enumerates all its possible triggers of size $s \times s$ (Lines 3–10). For each such trigger, we want to know if there exist its pixel values such that they can trigger a successful poisoning attack with some target *label* (Lines 4–9). Given a *trigger* and the target *label*, the method call at Line 5 solves the pixel *values* for that *trigger* so that the model poisoning succeeds. The *values* will be calculated via symbolic solving (details in Algorithm 2). It can happen that there do not exist any values of pixels in *trigger* that could lead samples in $\mathcal{T}$ to be poisoned and classified as the target *label*. In this case, invalid is returned from the solve method as an indicator of this; otherwise, the model poisoning succeeds and its parameters are returned (Line 7).

In VPN, given an input x in $\mathcal{T}$, if all its possible triggers have been tested against all possible labels and there is no valid poisoning *values*, then n_unsat is incremented by 1 (Line 11) for recording the un-poison-able inputs. Note that, for a successful model poisoning, it is not necessary all samples in $\mathcal{T}$ are successfully poisoned, as long as the number of poisoning misses is bounded by k. Therefore, a variable n_unsat is declared (Line 1) to record the number of samples in $\mathcal{T}$ from which a trigger cannot be found for a successful poisoning attack. If this counter (i.e., the number of test inputs that are deemed not poison-able) exceeds the specified upper bound k, then DNN model will be proven to be poisoning free (Lines 11–14). Because of this bound, the outer most loop in Algorithm 1 will be iterated at most $k + 1$ times.

Constraint Solving Per Trigger-Label Combination. In Algorithm 2, the method *solve_trigger_for_label* searches for valid pixel values of *trigger* such that not only the input x is classified by the DNN f as the target *label* after assigning these values to the *trigger*, but also this generalizes to other inputs in the test set $\mathcal{T}$, subject to maximum poisoning misses k.

The major part of Algorithm 2 is a while loop (Lines 3–15). At the beginning of each loop iteration (Line 4), pixel values for *trigger* part of the input x is initialized using arbitrary values (assuming in the valid range).

Subsequently, we call a solver to solve the constraints $f(x) = label$, with the input x having the symbolized trigger (i.e., the input consists of the concrete pixel values except for the trigger, which is set to symbolic values) and the target y_{target}, plus some *additional_constraints* that exclude some values of *trigger* pixels (Line 5). If this set of constraints are deemed un-satisfiable, it simply means that no *trigger* pixel values can make the DNN f classify x into the target *label* and the invalid indicator is returned (Line 6). Otherwise, at Line 8, we call the solver to get the *values* that satisfy the *if* constraints set at Line 5.

Algorithm 2. *solve_trigger_for_label*

INPUT: DNN f, test set $\mathcal{T}$, poisoning misses k, image x, *trigger*, target *label*
OUTPUT: pixel *values* for *trigger*

1: *additional_constraints* $\leftarrow \{\}$
2: *values* $\leftarrow$ invalid
3: **while** *values* = invalid and early termination condition is not met **do**
4:     $x[patch] \leftarrow symbolic\_non\_deterministic\_variables()$
5:     **if** *solver.solve*$(\{f(x) = label\} \cup additional\_constraints) =$ unsat **then**
6:         **return** invalid
7:     **end if**
8:     *values* $\leftarrow$ *solver.get_solution()*
9:     **if** $(trigger, values)$ and *label* satisfy Eq. (2) for $\mathcal{T}$, k **then**
10:         **return** *values*
11:     **else**
12:         *additional_constraints* $\leftarrow$ *additional_constraints* $\cup \{x[trigger] \neq values\}$
13:         *values* $\leftarrow$ invalid
14:     **end if**
15: **end while**
16: **return** invalid

We do not assume any specific solver or DNN verification tool. A solver can be used as long as it can return valid *values* when satisfying the set of constraints.

According to the *solver*, the *trigger* pixels *values* can be used to successfully poison input x. At this stage, we still need to check if it enables successful poisoning attack on other inputs in the test suite $\mathcal{T}$. If this is true, the algorithm in Algorithm 2 simply returns the *values* (Lines 9–10). Otherwise, the while loop will continue. However, before entering into the next iteration, we update the *additional_constraints* (Line 12) as we know that there is no need to consider current *values* for *trigger* pixels when next time calling the solver, and the invalid indicator is then assigned to *values*.

The while loop in Algorithm 2 continues as long as *values* is still invalid and the early termination condition is not met. The early termination condition can be e.g., runtime limit. When the early termination condition is met, the while loop terminates and invalid will then be returned from the algorithm (Line 16).

Correctness and Termination. Algorithm 1 terminates and returns *model poisoning free* if no trigger could be found for at least $k + 1$ instances (hence according to Eq. 2 the model is not poisoned). Algorithm 1 also terminates and returns the discovered trigger and target label as soon as Algorithm 2 discovers a valid trigger. The trigger returned by Algorithm 2 is valid as it satisfies Eq. (2) (lines 9–10).

2.3   Achieving Scalability via Attack Transferability

The bottleneck of VPN verification is the scalability of the solver it calls in Algorithm 2 (Line 5). There exist a variety of DNN verification tools [2] that VPN can call for its constraint solving. However, there is an upper bound limit on the DNN model complexity for such tools to handle. Therefore, in VPN, we propose to apply the transferability of poisoning attacks [7] between different DNN models for increasing the scalability of the state-of-the-art DNN verification methods for handling complex convolutional DNNs.

Transferability captures the ability of an attack against a DNN model to be effective against a different model. Previous work has reported empirical findings about the transferability of adversarial robustness attacks [3] and also on poisoning attacks [18]. VPN smartly uses this transferability for improving its scalability.

Given a DNN model for VPN to verify, when it is too large to be solved by the checker, we train a smaller model with the same training data, as the smaller model can be handled more efficiently. Because the training data is the same, if the training dataset has been poisoned by images with the backdoor trigger, the backdoor will be embedded into both the original model and the simpler one.

Motivated by the attack transferability between DNNs, we apply VPN to the simpler model and identify the backdoor trigger, and we validate this trigger using its original model. Empirical results in the experiments (Sect. 3) show the effectiveness of this approach for identifying model poisoning via transferability.

Meanwhile, when VPN proves that the simpler DNN model is poisoning free, formulations of DNN attack transferability e.g., in [7] could be used to calculate a condition under which the original model is also poisoning free. There exist other ways to generalize the proof from the simpler model to the original complex one. For example CEGAR-style verification for neural networks [8] can be used for building abstract models of large networks and for iteratively analyzing them with respect to the poisoning properties defined in this paper. Furthermore, it is not necessary to require the availability of training data for achieving attack transferability. Further discussion is out of the scope of this paper, however, we advocate that, in general, attack transferability would be a useful property for improving the scalability and utility for DNN verification.

3   Evaluation

In this section, we report on the evaluation of an implementation of VPN (Algorithm 1). Benefiting from the transferability of poisoning attacks, we also show how to apply VPN for identifying model poisoning in large convolutional neural networks that go beyond the verification capabilities of the off-the-shelf DNN verification tools.

3.1   Setup

Datasets and DNN Models. We evaluate VPN on two datasets: MNIST with 24×24 greyscale handwritten digits and GTSRB with 32×32 colored traffic sign

Table 1. Poisoned models. 'Clean Accuracy' is each model's performance on its original test data, *which is not necessarily the same as the test set* T *in* VPN *algorithm.* 'Attack Success Rate' measures the percentage of poisoned inputs, by placing the trigger on original test data, that are classified as the target label.

Model	Clean Accuracy	Attack Success Rate	Model Architecture
MNIST-FC1	92.0%	99.9%	10 dense × 10 neurons
MNIST-FC2	95.0%	99.1%	10 dense × 20 neurons
MNIST-CONV1	97.8%	99.0%	2 conv + 2 dense (Total params: 75,242)
MNIST-CONV2	98.7%	98.9%	2 conv + 2 dense (Total params: 746,138)
GTSRB-CONV1	97.8%	100%	6 conv (Total params: 139,515)
GTSRB-CONV2	98.11%	100%	6 conv (Total params: 494,251)

images. Samples of the poisoned data are shown in Fig. 1. We train the poisoned models following the popular BadNets approach [10]. We insert the Firefox logo into GTSRB data using the TABOR tool in [11].

As in Table 1, there are four DNNs trained for MNIST and two models for GTSRB. The model architecture highlights the complexity of the model. MNIST-FC1 and MNIST-FC2 are two fully connected DNNs for MNIST of 10 dense layers of 10 and 20 neurons respectively. MNIST-CONV1 and MNIST-CONV2 are two convolutional models for MNIST. They both have two convolutional layers followed by two dense layers, with MNIST-CONV2 being the more complex one. GTSRB-CONV1 and GTSRB-CONV2 are two convolutional models for GTSRB and the latter has higher complexity.

Verification Tools. VPN does not require particular solvers and we use Marabou[2] and nneum[3] in its implementation. Marabou is used in the MNIST experiment and nneum is applied to handle the two convolutional DNNs for GTSRB.

3.2   Results on MNIST

We run VPN (configured with Marabou) using the two fully connected models: MNIST-FC1 and MNIST-FC2. We arbitrarily sample 16 input images to build the test suite T in the VPN algorithm. For testing purpose, we configure the poisoning missing tolerance number as $k = |T| - 1$, that is, whenever the constraints solver returns some valid trigger values, VPN stops. The early termination condition in Algorithm 2 is set up as a 1,800 s timeout. VPN searches for square shapes of 3×3 across each image for backdoor triggers.

Figure 2 shows several backdoor trigger examples found by VPN. We call them the synthesized triggers via VPN. Compared with the original trigger in Fig. 1, the synthesized ones do not necessarily have the same values or even the same positions. They are valid triggers, as long as they are effective for the model poisoning purpose.

---

[2] Github link: https://github.com/NeuralNetworkVerification/Marabou (commit number 54e76b2c027c79d56f14751013fd649c8673dc1b).

[3] Github link: https://github.com/stanleybak/nnenum (commit number fd07f2b6c55 ca46387954559f40992ae0c9b06b7).

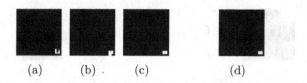

(a)   (b) .   (c)   (d)

Fig. 2. Synthesized backdoor triggers via VPN: (a)(b)(c) are from MNIST-FC1 and (d) is from MNIST-FC2. A trigger is solved as a bounded square. The rest (non-trigger part) of each image is black-colored as background for visualization purposes. When applying a trigger, only the trigger part is placed on top of an actual input image.

Table 2. Attack success rates across different models by the synthesized triggers via VPN (in Fig. 2). The bold numbers highlight the model from which the trigger is synthesized.

Synthesized trigger	MNIST-FC1	MNIST-FC2	MNIST-CONV1	MNIST-CONV2
Figure 2(a)	**95.7%**	85.8%	57.9%	39.9%
Figure 2(b)	**96.7%**	94.0%	74.5%	68.6%
Figure 2(c)	**96.7%**	93.7%	64.4%	80.1%
Figure 2(d)	97.3%	**94.7%**	70.2%	81.1%

Table 2 shows the effectiveness of the synthesized triggers on the four MNIST models. Thanks to the transferability property (discussed in Sect. 2.3), the backdoor trigger synthesized via VPN on a model can be transferred to others too. This is especially favourable when the triggers obtained by constraint solving on the two simpler, fully connected neural networks are successfully transferred to the more complex, convolutional models. Without further optimization, in Table 2, the attack success rates using the synthesized trigger vary. Nevertheless, it should be alarming enough when 39.9% (the lowest attack success rate observed) of the input images are classified as the target label '7'.

3.3 Results on GTSRB

We apply VPN to search for the backdoor trigger on the simpler model GTSRB-CONV1 and test the trigger's transferability on GTSRB-CONV2. T is the original GTSRB test set (excluding those wrongly classified tests) and $k = |T| - 1$.

The trigger found via VPN for GTSRB is shown in Fig. 3. It takes the solver engine nneum 5,108 s to return the trigger values. After using this synthesized trigger, more than 30% of images from GTSRB test dataset will be classified by GTSRB-CONV1 as the target label 'turn right' (out of the 43 output classes), which we believe is a high enough attack success rate for triggering model poisoning warning. Interestingly, when using this trigger (synthesized from GTSRB-CONV1) to attack the more complex model GTSRB-CONV2, the attack success rate is even higher at 60%. This observation motivates us to investigate in the future if there are conditions that triggers would affect more complex network architectures but not the simpler ones.

Fig. 3. Synthesized backdoor triggers via VPN from the poisoned model GTSRB-CONV1. The identified target label is 'turn right'.

Fig. 4. Synthesized backdoor trigger via VPN from the clean model MNIST-FC1-Clean. The identified target label is '2'.

3.4 Results on Clean Models

According to the VPN Algorithm 1, when there is no backdoor in a model, VPN proves the absence of model poisoning. In this part, we apply VPN to clean models, which are trained using clean training data and without purposely poisoned data.

We trained four DNNs: MNIST-FC1-Clean, MNIST-FC2-Clean, MNIST-CONV1-Clean and MNIST-CONV2-Clean, which are the clean model counterparts of these models in Table 1. All other setups are the same as the MNIST experiments in Sect. 3.2.

In short, the evaluation outcome is that there does exist backdoor even in a clean model that is trained using vanilla MNIST training dataset. Figure 4 shows one such trigger identified by VPN. It leads to 57.3% attack success rate for MNIST-FC1-Clean and 68.2% attack success rate for MNIST-FC2-Clean. Even though these rates on clean models are not as high as the attack success rates for these poisoned models, they are still substantially higher than the portion of input images with groundtruth label '2'.

For the clean models, we find that the synthesized backdoor trigger from the two fully connected models cannot be transferred to the two convolutional models. Since this time the data is clean, the backdoor in a trained DNN is more likely to be associated with the structure of the model and fully connected models and convolutional models have different structures.

4   Conclusion

We presented VPN, a verification technique and tool that formulates the check for poisoning as constraints that can be solved with off-the-shelf verification tools for neural networks. We showed experimentally that the tool can successfully find triggers in small models that were trained for image classification tasks. Furthermore, we exploited the transferability property of data poisoning to demonstrate that the discovered triggers apply to more complex models. Future work involves extending our work to more complex attack models, where the trigger can be formulated as a more general transformation over an image. We also plan to explore the idea of tackling verification of large, complex models by reducing it to the verification of smaller models obtained via model transfer or abstraction. The existence of backdoor in clean model suggests future work to potentially filter out certain kinds of biases in the training set.

References

1. Bak, S.: nnenum: verification of relu neural networks with optimized abstraction refinement. In: Dutle, A., Moscato, M.M., Titolo, L., Muñoz, C.A., Perez, I. (eds.) NFM 2021. LNCS, vol. 12673, pp. 19–36. Springer, Cham (2021). https://doi.org/10.1007/978-3-030-76384-8_2

2. Bak, S., Liu, C., Johnson, T.: The second international verification of neural networks competition (vnn-comp 2021): summary and results. arXiv preprint arXiv:2109.00498 (2021)

3. Biggio, B., et al.: Evasion attacks against machine learning at test time. In: Blockeel, H., Kersting, K., Nijssen, S., Železný, F. (eds.) ECML PKDD 2013. LNCS (LNAI), vol. 8190, pp. 387–402. Springer, Heidelberg (2013). https://doi.org/10.1007/978-3-642-40994-3_25

4. Brown, T.B., Mané, D., Roy, A., Abadi, M., Gilmer, J.: Adversarial patch (2017). 10.48550/ARXIV.1712.09665. https://arxiv.org/abs/1712.09665

5. Cheng, S., Liu, Y., Ma, S., Zhang, X.: Deep feature space trojan attack of neural networks by controlled detoxification. In: Proceedings of the AAAI Conference on Artificial Intelligence, vol. 35, pp. 1148–1156 (2021)

6. Chiang, P.Y., Ni, R., Abdelkader, A., Zhu, C., Studor, C., Goldstein, T.: Certified defenses for adversarial patches. In: International Conference on Learning Representations (2019)

7. Demontis, A., et al.: Why do adversarial attacks transfer? explaining transferability of evasion and poisoning attacks. In: USENIX Security, pp. 321–338 (2019)

8. Elboher, Y.Y., Gottschlich, J., Katz, G.: An abstraction-based framework for neural network verification. In: Lahiri, S.K., Wang, C. (eds.) CAV 2020. LNCS, vol. 12224, pp. 43–65. Springer, Cham (2020). https://doi.org/10.1007/978-3-030-53288-8_3

9. Gao, Y., Xu, C., Wang, D., Chen, S., Ranasinghe, D.C., Nepal, S.: Strip: A defence against trojan attacks on deep neural networks. In: Proceedings of the 35th Annual Computer Security Applications Conference, pp. 113–125 (2019)

10. Gu, T., Liu, K., Dolan-Gavitt, B., Garg, S.: Badnets: evaluating backdooring attacks on deep neural networks. IEEE Access 7, 47230–47244 (2019). https://doi.org/10.1109/ACCESS.2019.2909068

11. Guo, W., Wang, L., Xing, X., Du, M., Song, D.: Tabor: a highly accurate approach to inspecting and restoring trojan backdoors in AI systems. arXiv preprint arXiv:1908.01763 (2019)

12. Houben, S., Stallkamp, J., Salmen, J., Schlipsing, M., Igel, C.: Detection of traffic signs in real-world images: the german traffic sign detection benchmark. In: International Joint Conference on Neural Networks, No. 1288 (2013)

13. Huang, X., et al: A survey of safety and trustworthiness of deep neural networks: verification, testing, adversarial attack and defence, and interpretability. Comput. Sci. Rev. 37, 100270 (2020)

14. Katz, Guy, Katz, G.: The Marabou Framework for Verification and Analysis of Deep Neural Networks. In: Dillig, Isil, Tasiran, Serdar (eds.) CAV 2019. LNCS, vol. 11561, pp. 443–452. Springer, Cham (2019). https://doi.org/10.1007/978-3-030-25540-4_26

15. LeCun, Y., Bottou, L., Bengio, Y., Haffner, P.: Gradient-based learning applied to document recognition. Proc. IEEE 86(11), 2278–2324 (1998)

16. Liu, Y., et al.: Trojaning attack on neural networks. In: NDSS (2018)

17. Steinhardt, J., Koh, P.W.W., Liang, P.S.: Certified defenses for data poisoning attacks. Advances in neural information processing systems 30 (2017)
18. Suciu, O., Marginean, R., Kaya, Y., Daume III, H., Dumitras, T.: When does machine learning {FAIL}? generalized transferability for evasion and poisoning attacks. In: USENIX Security (2018)
19. Usman, M. et al.: NNREPAIR: constraint-based repair of neural network classifiers. Computer Aided Verification , 3–25 (2021). https://doi.org/10.1007/978-3-030-81685-8_1
20. Wang, B., et al.: Neural cleanse: identifying and mitigating backdoor attacks in neural networks. In: 2019 IEEE Symposium on Security and Privacy (SP), pp. 707–723. IEEE (2019)

A Cascade of Checkers for Run-time Certification of Local Robustness

Ravi Mangal[1(✉)] and Corina Păsăreanu[1,2]

[1] Carnegie Mellon University CyLab, Pittsburgh, PA 15213, USA
{rmangal,pcorina}@andrew.cmu.edu
[2] NASA Ames, Moffett Field, CA 94035, USA

Abstract. Neural networks are known to be susceptible to adversarial examples. Different techniques have been proposed in the literature to address the problem, ranging from adversarial training with robustness guarantees to post-training and run-time certification of local robustness using either inexpensive but incomplete verification or sound, complete, but expensive constraint solving. We advocate for the use of a run-time cascade of over-approximate, under-approximate, and exact local robustness checkers. The exact check in the cascade ensures that no unnecessary alarms are raised, an important requirement for autonomous systems where resorting to fail-safe mechanisms is highly undesirable. Though exact checks are expensive, via two case studies, we demonstrate that the exact check in a cascade is rarely invoked in practice. Code and data are available at https://github.com/ravimangal/cascade-robustness-impl.

Keywords: Neural networks · Local robustness · Run-time checks

1 Introduction

Software systems with neural network components are becoming increasingly common due to the new computational capabilities unlocked by neural networks. However, the susceptibility of neural network classifiers to adversarial examples is concerning, particularly for networks used in safety-critical systems. Adversarial examples [30] are inputs produced by applying small, imperceptible modifications to correctly classified inputs such that the modified input is classified incorrectly. This lack of *robustness* of neural networks to small changes in the inputs can not only be exploited by malicious actors [4,30] but also lead to incorrect behavior in the presence of natural noise [12]. Moreover, this phenomenon is widespread - neural networks trained for a variety of tasks like image recognition [4,30], natural language processing [1,14], and speech recognition [5,6,25] have been shown to exhibit the same weakness.

Recognizing the seriousness of the problem, the research community has been actively engaged in studying it. We now know that a neural classifier does not only need to be accurate (i.e., make correct predictions) but it also needs to be *locally robust* at all inputs of interest. A network is ϵ-locally robust at an input

O. Isac et al. (Eds.): NSV 2022/FoMLAS 2022, LNCS 13466, pp. 15–28, 2022.
https://doi.org/10.1007/978-3-031-21222-2_2

x if it makes the same prediction at all the inputs that lie within a ball of radius ϵ centered at x (assuming some distance metric defined over the input domain). Local robustness at an input ensures that small ($\leq \epsilon$) modifications to the input do not affect the network prediction.

In practice, ensuring that neural networks are robust has turned out to be a hard challenge. A number of approaches modify the neural network training process to encourage learning robust networks [19, 20, 26, 34]. While such robustness-aware training can greatly reduce the susceptibility of neural networks to adversarial examples due to enhanced robustness, they provide no guarantee that the trained network is free from adversarial examples. For safety-critical applications, the existence of a single adversarial example can have disastrous consequences.

How can we ensure that a network is free from adversarial examples? One approach is to check, using a local robustness certification procedure, if a network is locally robust at every input of interest. This requires assuming that all inputs of interest are known a priori, an assumption that is unlikely to hold in practice. The only option then, to guarantee protection from adversarial examples, is to check at run-time if the network is locally robust at the evaluated input x. If the run-time check passes, we are assured that the input x cannot be adversarial (since even if x is an ϵ-perturbed input produced by an adversary, local robustness at x ensures that the network assigns it the same label as the original unperturbed input). If the check fails, then x is potentially an adversarial input, and one has to resort to some fail-safe mechanism like aborting execution or asking a human expert to make the prediction, both undesirable scenarios to be avoided as far as possible. Though a run-time check introduces a computational overhead, it is the only mechanism for ensuring that a safety-critical system does not misbehave due to adversarial examples.

While the problem of checking if a neural network is locally robust at an input is known to be NP-Complete for ReLU neural networks [15], a number of practical algorithms that variously balance the trade-off between precision of the check and efficiency have been proposed in the literature. Sound but incomplete (or *over-approximate*) algorithms guarantee that the network is locally robust whenever the check passes but not vice versa, i.e., they can report false positives [10, 11, 19, 26, 29, 33, 34]. Sound and complete (or *exact*) algorithms are guaranteed to either find a valid robustness certificate or a valid counterexample but they can be very inefficient [15, 31]. Attack (or bug-finding or *under-approximate*) algorithms only aim to find counterexamples to robustness but can fail to find a counterexample even if one exists (and are therefore unable to provide a robustness certificate) [2, 4, 8, 20, 24, 32].

When deploying local robustness checks at run-time, a common choice is to use over-approximate algorithms because of their computationally efficient nature [19]. But, due to their incompleteness, these algorithms can report false positives and unnecessarily require the use of the undesirable fail-safe mechanisms. On the other hand, while an exact check can avoid unnecessary alarms, these checks involve constraint solving and can be prohibitively expensive. In order to balance between the frequency of unnecessary alarms and the cost of a local robustness check, in this paper, we propose to use a cascade of local robustness checkers at run-time. In particular, we propose a cascade of checks that starts with an over-approximate check, followed by an under-approximate check (which is also computationally cheap), and ends with an exact check. This sequence ensures that if the first check fails, we first attempt to evaluate if the failure was due to a false positive or a true positive. If the under-approximate check (i.e., the attack algorithm) succeeds, then it was indeed a true positive and we are forced to resort to our undesirable fail-safe. However, if the attack also fails, we use the exact check that either returns a certificate of robustness or a counterexample. Notice that the exact check, which is computationally expensive, is invoked only if absolutely necessary. Though a local robustness check that combines an over-approximate algorithm with an exact algorithm has been proposed before [28], our approach differs in multiple ways. Most importantly, while the check from [28] closely integrates the over-approximate and exact algorithms, our approach is entirely agnostic to the internal implementation details of the checks being composed.

We have implemented our run-time cascade of checkers, and empirically evaluate it using two real-world case studies. Our evaluation suggests that, in practice, a cascaded checker can be a reasonable choice since the over-approximate and under-approximate checks are able to resolve most of the local robustness queries, and the expensive, exact check is rarely invoked.

The rest of the paper is organized as follows. In Sect. 2, we provide the necessary preliminaries and definitions. In Sect. 3, we briefly describe techniques for robustness-aware training of neural networks as well as the algorithms used for checking local robustness. In Sect. 4, we give more details about our run-time cascade of local robustness checkers. In Sect. 5, we present our two case studies and empirically evaluate our run-time cascade of checks. Finally, we conclude in Sect. 6.

2   Background

We present preliminaries and necessary definitions in this section.

Neural Networks. A neural network, $f_\theta : \mathbb{R}^d \to \mathbb{R}^m$, is a function defined by a composition of linear and non-linear transformations, where θ refers to *weights* or parameters characterizing the linear transformations. As the internal details of neural networks are not relevant to much of this paper, we will by default omit the subscript θ, and treat f as a black-box function. Neural networks are used as classifiers by extracting *class predictions* from the output $f(x) : \mathbb{R}^m$, also called the *logits* of a network. Given a neural network f, we use the uppercase F to refer to the corresponding neural classifier that returns the top class: $F = \lambda x.\,\mathrm{argmax}_i\, f_i(x)$. For our purposes, we will assume that argmax returns a single index, $i^* \in [m]^1$; ties may be broken arbitrarily

Adversarial Examples. An adversarial example for a neural classifier is the result of applying small modifications to a correctly classified *valid* input such that the modified input is classified incorrectly. Definition 1 below formalizes the notion of an adversarial example.

Definition 1 (Adversarial Example). *Given a neural classifier $F \in \mathbb{R}^d \rightarrow [m]$ and an input $x \in \mathbb{R}^d$, an input x' is an adversarial example with respect to an ℓ_p distance metric, and a fixed constant $\epsilon \in \mathbb{R}$ if*

$$||x - x'||_p \leq \epsilon \wedge F(x) \neq F(x')$$

Local Robustness. A classifier is protected from adversarial examples with respect to a valid input x if it is locally robust at x. As stated in Definition 2, a classifier is locally robust at x, if its prediction does not change in an ϵ-ball centered at x.

Definition 2 (Local Robustness). *A neural classifier $F \in \mathbb{R}^d \rightarrow [m]$ is (ϵ, ℓ_p)-locally robust at $x \in \mathbb{R}^d$ if,*

$$\forall x' \in \mathbb{R}^d. \ ||x - x'||_p \leq \epsilon \implies F(x') = F(x)$$

Here we consider robustness and input modifications with respect to l_p norms, commonly used in the literature, but our approach extends also to other modifications, which are not necessarily captured with l_p norms.

Certified Run-time Defense Against Adversarial Examples. Before deployment, we can evaluate the susceptibility of a trained neural classifier to adversarial examples by checking its local robustness at inputs in the training and test datasets. However, this provides no guarantee of immunity from adversarial examples on unseen inputs. Checking local robustness of a classifier during run-time can provide such a guarantee. If the classifier is locally robust at the input x under evaluation, then x cannot be an adversarial example. Even if x is an ϵ-perturbed input generated by an adversary, the local robustness certificate ensures that the classifier's prediction is not affected by the perturbation. However, if the local robustness check fails, to be safe, one has to assume that x is potentially an adversarial example that can cause the classifier to misbehave, and resort to using a fail-safe mechanism (like aborting execution or deferring to a human expert) designed to handle this scenario. Defending against adversarial examples at run-time via local robustness checks is a well-known technique [7,19].

---

[1] $[m] := \{0, \ldots, m-1\}$.

3   Ensuring Local Robustness

There are two primary approaches to ensure that neural classifiers are locally robust. One approach is to train the classifiers in a robustness-aware manner. The other approach is to check for local robustness violations by the classifier at run-time. In this section, we provide a brief overview of techniques for robustness-aware training of classifiers as well as algorithms for checking local robustness.

3.1   Local Robustness via Training

Standard training of neural classifiers is typically framed as the following optimization problem:

$$\theta^{\star} = \underset{\theta}{\operatorname{argmin}} \sum_{i=1}^{i=n} L(F_\theta(x_i), y_i) \tag{1}$$

where F is a family of neural networks parameterized by $\theta \in \mathbb{R}^p$, $(x_i, y_i) \in \mathbb{R}^d \times [m]$ is a labeled training sample, and L is a real-valued *loss function* that measures how well F_θ "fits" the training data. $F_{\theta^{\star}}$ is the final trained model. To train models in a robustness-aware manner, the optimization objective is modified in order to promote both accuracy and local robustness of the trained models. A very popular robustness-aware training approach, first proposed by [20], formulates the following min-max optimization problem:

$$\theta^{\star} = \underset{\theta}{\operatorname{argmin}} \sum_{i=1}^{i=n} \max_{\delta \in \mathbb{B}(0,\epsilon)} L(F_\theta(x_i + \delta), y_i) \tag{2}$$

where $\mathbb{B}(0, \epsilon)$ is a ball of radius ϵ centered at 0. Intuitively, this optimization objective captures the idea that we want classifiers that perform well even on adversarially perturbed inputs. The inner maximization problem aims to find the worst-case adversarially perturbed version of a given input that maximizes the loss, formalizing the notion of an adversary attacking the neural network. On the other hand, the outer minimization problem aims to find classifier parameters so that the worst-case loss given by the inner optimization problem is minimized, capturing the idea that the trained classifier needs to be immune to adversarial perturbations. Solving the optimization problem in Eq. 2 can be very computationally expensive, and most practically successful algorithms only compute approximate solutions to the inner maximization problem. They either compute an under-approximation (lower bound) of the maximum loss [20] or compute an over-approximation (upper bound) of the maximum loss [21,26,34].

An alternate formulation of the optimization objective for robustness-aware training is as follows:

$$\theta^{\star} = \underset{\theta}{\operatorname{argmin}} \sum_{i=1}^{i=n} (L(F_\theta(x_i), y_i) + L_{rob}(F_\theta(x_i), \epsilon)) \tag{3}$$

where L_{rob} measures the degree to which classifier F_θ is ϵ-locally robust at x_i. Approaches using an optimization objective of this form [19,36] are required to check local robustness of the neural classifier in order to calculate L_{rob}. As long as the local robustness checking procedure is differentiable, any such procedure can be used. For our case studies (Sect. 5), we use an approach based on Eq. 3, namely GloRo Nets [19], to train our classifiers. To calculate L_{rob}, GloRo uses a sound but incomplete check for local robustness based on calculating the Lipschitz constant of the neural network f.

3.2 Run-time Checks for Local Robustness

Algorithms that check if neural classifiers are locally robust come in three primary flavors:

1. *over-approximate* (or sound but complete) algorithms that guarantee local robustness of the neural network whenever the check passes but not vice versa, i.e., the check can fail even if the network is locally robust,
2. *under-approximate* (or attack) algorithms that generate counterexamples highlighting violations of local robustness but are not always guaranteed to find counterexamples even when they exist,
3. *exact* (sound and complete) algorithms that are guaranteed to either find a certificate of local robustness or a counterexample, but can be very computationally expensive.

Not only are such algorithms useful for checking local robustness at run-time, but they are also useful for evaluating the quality of the trained neural network pre-deployment. In particular, these checkers can be used to evaluate the local robustness of the neural network at the inputs in the training and test datasets. If the trained network lacks local robustness on these known inputs, it is unlikely to be locally robust on unknown inputs. We briefly survey local robustness checking algorithms in the rest of this section.

Over-Approximate Algorithms. A variety of approaches have been used for implementing over-approximate algorithms for checking local robustness. Algorithms using abstract interpretation [11,29] approximate the ϵ-ball in the input space with a polyhedron enclosing the ball and symbolically propagate the polyhedron through the neural network to get an over-approximation of all the possible outputs of the network when evaluated on points in the input polyhedron. This information can then be used to certify if the network is locally robust or not. Other algorithms frame local robustness certification as an optimization problem [3,9,26,31,34]. A key step in these algorithms is to translate the neural network into optimization constraints. This translation, if done in a semantics-preserving manner, can lead to intractable optimization problems. Instead, the translation constructs relaxed constraints that can be solved efficiently, at the cost of incompleteness. Another approach for over-approximate certification of local robustness relies on computing local or global Lipschitz constant for the neural network under consideration. The Lipschitz constant of a function upper bounds the rate of change of the function output with respect to its input. Given the Lipschitz constant of a neural network and its logit values at a particular input x, one can compute a lower bound of the radius of the ball centered at x within which the network prediction does not change. If this lower bound is greater than ϵ, then we have a certificate of ϵ-local robustness of the network at x. A number of certification algorithms are based on computing the Lipschitz constant [19,33], and we use one such approach [19] in our case studies in Sect. 5.

Under-Approximate Algorithms. Under-approximate algorithms, usually referred to as *attacks* in the adversarial machine learning community, can be divided into two major categories. White-box algorithms [4,13,20] assume that they have access to the internals of the neural network, namely the architecture and the weights of the network. Given such access, these algorithms frame the problem of finding counterexamples to local robustness for classifier F at input x as an optimization problem of the form,

$$\delta^{\star} = \underset{\delta \in \mathbb{B}(0, \epsilon)}{\operatorname{argmax}} L(F(x + \delta), F(x)) \tag{4}$$

The counterexample is given by $x + \delta^{\star}$. Intuitively, the algorithms try to find a perturbed input $x' := x + \delta^{\star}$ such that x' is in the ϵ-ball centered at x, and the classifier output at x' is as different from x as possible (formalized by the requirement to maximize the loss L). This optimization problem is non-convex since F can be an arbitrarily complicated function, and in practice, attack algorithms use gradient ascent to solve the optimization problem. Due to the non-convex nature of the optimization objective, such algorithms are not guaranteed to find the optimal solution, and therefore, are not guaranteed to find a counterexample even if one exists. Black-box algorithms [24] are the other category of attack algorithms, and such algorithms only assume query access to the neural network, i.e., the algorithms can only observe the network's outputs on queried inputs, but do not have access to the weights and therefore, cannot directly access the gradients. In other words, black-box algorithms assume a weaker adversary than white-box algorithms. For our case studies, we use a white-box attack algorithm [20].

Exact Algorithms. Exact algorithms for checking local robustness [15,16,31] encode a neural network's semantics as system of semantics-preserving constraints, and pose local robustness certification as constraint satisfaction problems. Though these algorithms are guaranteed to either find a certificate of robustness or a counterexample, they can be quite computationally expensive. An alternate approach for exact checking constructs a smoothed classifier from a given neural classifier using a randomized procedure at run-time [7,18,27,35]. Importantly, the smoothed classifier is such that, at each input, its local robustness radius can be calculated exactly using a closed-form expression involving the outputs of the smoothed classifier. However, the randomized smoothing procedure can be very expensive as it requires evaluating the original classifier on a

Table 1. Trade-offs made by the different flavors of local robustness checkers.

	Finds certificates?	Finds counterexamples?	Efficient?
Over-approximate	✓	×	✓
Under-approximate	×	✓	✓
Exact	✓	✓	×

Algorithm 4.1: Cascaded local robustness checker

Inputs: A neural classifier $F \in \mathbb{R}^d \to [m]$, an input $x \in \mathbb{R}^d$, and local
robustness radius $\epsilon \in \mathbb{R}$
Output: Certified status $b \in \{1, 0\}$

```
1  Cascade(F , x , epsilon):
2      cert := Over-approximate(F, x, ε)
3      if ¬cert then
4          cex := Under-approximate(F, x, ε)
5          if ¬cex then
6              b := Exact(F, x, ε)
7          else
8              b := 0
9      else
10         b := 1
11     return b
```

---

large number of randomly drawn samples. For our case studies, we use an exact
algorithm based on constraint-solving.

4   A Cascade of Run-time Local Robustness Checkers

The previous section demonstrates that one has many options when picking a
local robustness checker to be deployed at run-time. Every option offers a different trade-off between the ability to certify local robustness and the efficiency of
the check. Table 1 summarizes these trade-offs. For each flavor of local robustness
checkers, the table shows if the checkers are able to produce certificates of local
robustness, find counterexamples, and do so efficiently. Ideally, we would like a
checker to possess all these characteristics but the NP-Complete nature of the
problem makes this impossible. In light of these trade-offs, a common choice is
to deploy an over-approximate checker [19]. Such checkers can falsely report that
the neural network is not locally robust, causing unnecessary use of the fail-safe
mechanisms.

We propose a local robustness checker that combines existing local robustness
checkers of different flavors in a manner that brings together their strengths

while mitigating their weaknesses. Our *cascaded* checker uses a sequence of over-approximate, under-approximate, and exact checkers. Algorithm 4.1 describes the cascaded checker. The inputs to the algorithm are the neural classifier F, the input x where we want to certify local robustness of F, and the radius ϵ to be certified. The algorithm first invokes an `Over-approximate` checker (line 2). If F is certified to be locally robust at x (i.e., `certs` equals 1), then we are done. Otherwise, the algorithm calls an `Under-approximate` checker (lines 3–4). If the under-approximate checker succeeds in finding a counterexample (i.e., `cex` equals 1), then we are done and know that F is not locally robust at x. Otherwise, an `Exact` checker is invoked (lines 5–6). The `Exact` checker either finds a proof of robustness ($b = 1$) or a counterexample ($b = 0$).

The cost of Algorithm 4.1, amortized over all the inputs seen at run-time, depends on the rate at with which the `Over-approximate` and `Under-approximate` checks succeed. If the cascaded checker has to frequently invoke the `Exact` checker, then one might as well directly use the `Exact` checker instead of the cascade. In practice, however, our empirical evaluation suggests that the `Exact` is rarely invoked (see Sect. 5). As a consequence, the cascaded checker is guaranteed to be sound and complete without incurring the high computational cost of an `Exact` checker.

5 Case Studies

The practical effectiveness of our sound and complete cascaded local robustness checker primarily depends on the run-time overhead introduced by the checker. This overhead, in turn, depends on the success rate of the `Over-approximate` and the `Under-approximate` checkers in the cascade. Given that the `Exact` checker is significantly more computationally expensive than the other components of the cascade, it is essential that it only be invoked rarely to ensure that the average overhead of the cascade per input is low.

We conduct two case studies to evaluate the rate at which the `Over-approximate` and `Under-approximate` checkers succeed. In particular, we measure the percentage of test inputs that are resolved by the `Over-approximate`, the `Under-approximate`, and the `Exact` checks. For both the case studies, we are interested in local robustness with respect to the ℓ_2 distance metric. Moreover, we train the neural classifiers in a robustness-aware manner using the state-of-the-art GloRo Net framework [19] that updates the loss function in the manner described in Eq. 3, and calculates the Lipschitz constant of the neural network in order to verify local robustness at an input. We also use this local robustness check based on Lipschitz constant as the `Over-approximate` check. For the `Under-approximate` check, we use the projected gradient descent (PGD) algorithm [20], as implemented in the CleverHans framework [23]. Finally, for the `Exact` check, we use the Marabou framework for neural network verification [16]. However, Marabou can only encode linear constraints, and so is unable to encode the ℓ_2 local robustness constraint. Instead, we use Marabou to check local robustness in the largest box contained inside the ℓ_2 ball of radius ϵ, and

Table 2. Percentage of inputs successfully handled by each check for MNIST.

Total queries	ϵ	% certified by Over-approximate	% attacked by Under-approximate	% resolved by Exact	% unresolved
10000	0.3	92.11	7.12	0.76	0.01
	1.58	45.02	49.52	5.28	0.18

in the smallest box containing the ϵ-ℓ_2 ball. If Marabou finds a counterexample for the first query, then we have a valid counterexample. Similarly, if Marabou is able to certify the second query, then we have a valid certificate. Though Marabou is no longer guaranteed to be sound and complete when used in the manner described, in our case studies, Marabou rarely fails to resolve the local robustness at an input.

5.1 MNIST

Our first case study uses the popular MNIST dataset [17] where the goal is to construct a neural classifier that can classify hand-written digits. Our neural network has three dense hidden layers with 64, 32, and 16 ReLU neurons in that order. We check ℓ_2 local robustness for an ϵ values of 0.3 and 1.58.

Table 2 shows the success rate of each of the local robustness checkers in our cascade. "Total queries" refers to the number of inputs in the test set used for evaluation. For ϵ value of 0.3, we see that the classifier is certified locally robust at 92.11% of the inputs by the Over-approximate check. Of the remaining 7.89% inputs, the Under-approximate check is able to find a counterexample for 7.12% of the inputs. As a result, only 0.77% of the 10000 inputs need to be checked with the Exact solver. Marabou is able to resolve 0.76% of the inputs, finding a counterexample in each case. Only 0.01% of the inputs, i.e., a single input, is not resolved by any of the checks (due to the fact that for ℓ_2 robustness queries, Marabou is not sound and complete). For ϵ value of 1.58, we see that the classifier is much less robust and the Over-approximate check is only able to certify 45.02% of the inputs. For all of the 5.28% inputs resolved by Marabou, it finds a counterexample. These results provide two interesting takeaways. First, the Exact checker is rarely invoked, suggesting that a cascaded checker is a reasonable choice in practice. Second, an Exact checker like Marabou is not only useful for finding certificates but also counterexamples.

5.2 SafeSCAD

Our second case study uses datasets produced as a part of the SafeSCAD[2] project. The project is concerned with the development of a driver attentiveness management system to support safe shared control of autonomous vehicles. Shared-control autonomous vehicles are designed to operate autonomously but

---

[2] Safety of Shared Control in Autonomous Driving.

Table 3. Percentage of inputs successfully handled by each check for SafeSCAD.

Total queries	ϵ	% certified by Over-approximate	% attacked by Under-approximate	% resolved by Exact	% unresolved
11819	0.05	54.36	35.77	8.49	1.38
	0.15	42.19	45.68	10.9	1.23

can request the driver to take over control if the vehicle enters conditions that the autonomous system cannot handle. The goal of the driver attentiveness management system then is to ensure that drivers are alert enough to take over control whenever requested. This system uses a neural network for predicting the driver alertness levels based on inputs from specialized sensors that monitor key vehicle parameters (velocity, lane position, etc.) and driver's biometrics (eye movement, heart rate, etc.). We used driver data collected as part of a SafeSCAD user study carried out within a driving simulator [22]. Our neural network has four dense hidden layers with 50, 100, 35, and 11 ReLU neurons in that order. We check ℓ_2 local robustness for ϵ values of 0.05 and 0.15.

Table 3 shows the success rate of each of the local robustness checkers in our cascade. We see that the trained classifier is not as robust as the MNIST case, and, for ϵ value of 0.05, it is certified locally robust only at 54.36% of the inputs by the Over-approximate check. Of the remaining 45.64% inputs, the Under-approximate check is able to find a counterexample for 35.77% of the inputs. 9.87% of the 11819 inputs need to be checked with the Exact solver. Marabou is able to resolve 8.49% of the inputs, finding a counterexample for 8.47% of the inputs and finding a proof of robustness for 0.02% of the inputs. 1.38% of the inputs not resolved by any of the checks (due to the fact that for ℓ_2 robustness queries, Marabou is not sound and complete). The results for ϵ value of 0.15 can be read off from the table in a similar manner. Note that Marabou finds a counterexample for all of the 10.9% of the inputs resolved by it. These results largely mirror the findings from the MNIST case study. In particular, they show that even when the neural classifier trained in a robustness-aware manner is not locally robust on a large percentage of the test inputs, the Over-approximate and Under-approximate checkers are able to resolve most of the inputs, and the Exact solver is rarely invoked.

6 Conclusion

In this paper, we surveyed techniques for checking local robustness on neural networks and we advocated for a cascade of checkers that best leverages their strengths and mitigates their weaknesses. We demonstrated the cascade of checkers with two case studies. Our experiments demonstrate that the most expensive check (which involves formal methods) is seldom needed as the previous checkers in the cascade are often sufficient for providing a robustness guarantee or for finding a counterexample. Nevertheless, the expensive, formal methods check is still important when dealing with autonomous, safety-critical systems as it can help avoid unnecessarily resorting to the fail-safe mechanism. Furthermore, we show that the formal methods check is useful for not only providing a certificate but also for producing counterexamples which are hard to find with cheaper techniques. Future work involves experimenting with more case studies and applying cascades to reasoning about more natural perturbations that are not necessarily captured with l_p-bounded modifications.

Acknowledgments. This material is based upon work supported by DARPA GARD Contract HR00112020006 and UK's Assuring Autonomy International Programme.

References

1. Alzantot, M., Sharma, Y., Elgohary, A., Ho, B.J., Srivastava, M., Chang, K.W.: Generating natural language adversarial examples. In: Proceedings of the 2018 Conference on Empirical Methods in Natural Language Processing, pp. 2890–2896. Association for Computational Linguistics, Brussels, Belgium (2018)
2. Athalye, A., Carlini, N., Wagner, D.: Obfuscated gradients give a false sense of security: circumventing defenses to adversarial examples. In: International Conference on Machine Learning, pp. 274–283. PMLR (2018)
3. Bastani, O., Ioannou, Y., Lampropoulos, L., Vytiniotis, D., Nori, A.V., Criminisi, A.: Measuring neural net robustness with constraints. In: Proceedings of the 30th International Conference on Neural Information Processing Systems, pp. 2621–2629. NIPS'16, Curran Associates Inc., Red Hook, NY, USA (2016)
4. Carlini, N., Wagner, D.: Towards evaluating the robustness of neural networks. In: 2017 IEEE Symposium on Security and Privacy (SP), pp. 39–57. IEEE Computer Society, Los Alamitos, CA, USA (2017). https://doi.org/10.1109/SP.2017.49,https://doi.ieeecomputersociety.org/10.1109/SP.2017.49
5. Carlini, N., et al.: Hidden voice commands. In: 25th USENIX Security Symposium (USENIX Security 16), pp. 513–530 (2016). https://www.usenix.org/conference/usenixsecurity16/technical-sessions/presentation/carlini
6. Carlini, N., Wagner, D.: Audio adversarial examples: targeted attacks on speech-to-text. In: 2018 IEEE Security and Privacy Workshops (SPW), pp. 1–7 (2018)
7. Cohen, J., Rosenfeld, E., Kolter, Z.: Certified adversarial robustness via randomized smoothing. In: Chaudhuri, K., Salakhutdinov, R. (eds.) In: Proceedings of the 36th International Conference on Machine Learning. Proceedings of Machine Learning Research, vol. 97, pp. 1310–1320. PMLR (2019). https://proceedings.mlr.press/v97/cohen19c.html
8. Croce, F., Hein, M.: Reliable evaluation of adversarial robustness with an ensemble of diverse parameter-free attacks. In: International Conference on Machine Learning, pp. 2206–2216. PMLR (2020)
9. Dvijotham, K., Stanforth, R., Gowal, S., Mann, T., Kohli, P.: A dual approach to scalable verification of deep networks. In: Proceedings of the Thirty-Fourth Conference Annual Conference on Uncertainty in Artificial Intelligence (UAI-18), pp. 162–171. AUAI Press, Corvallis, Oregon (2018)
10. Fromherz, A., Leino, K., Fredrikson, M., Parno, B., Păsăreanu, C.: Fast geometric projections for local robustness certification. In: International Conference on Learning Representations (ICLR) (2021)

11. Gehr, T., Mirman, M., Drachsler-Cohen, D., Tsankov, P., Chaudhuri, S., Vechev, M.: Ai2: Safety and robustness certification of neural networks with abstract interpretation. In: 2018 IEEE Symposium on Security and Privacy (SP), pp. 3–18 (2018)
12. Gilmer, J., Ford, N., Carlini, N., Cubuk, E.: Adversarial examples are a natural consequence of test error in noise. In: International Conference on Machine Learning, pp. 2280–2289. PMLR (2019)
13. Goodfellow, I.J., Shlens, J., Szegedy, C.: Explaining and harnessing adversarial examples. In: Bengio, Y., LeCun, Y. (eds.) In: 3rd International Conference on Learning Representations, ICLR 2015, San Diego, CA, USA, May 7–9, 2015, Conference Track Proceedings (2015). http://arxiv.org/abs/1412.6572
14. Jia, R., Liang, P.: Adversarial examples for evaluating reading comprehension systems. In: Proceedings of the 2017 Conference on Empirical Methods in Natural Language Processing, pp. 2021–2031. Association for Computational Linguistics, Copenhagen, Denmark (2017)
15. Katz, G., Barrett, C., Dill, D.L., Julian, K., Kochenderfer, M.J.: Reluplex: an efficient SMT solver for verifying deep neural networks. In: Majumdar, R., Kunčak, V. (eds.) CAV 2017. LNCS, vol. 10426, pp. 97–117. Springer, Cham (2017). https://doi.org/10.1007/978-3-319-63387-9_5
16. Katz, G., et al.: The marabou framework for verification and analysis of deep neural networks. In: Dillig, I., Tasiran, S. (eds.) CAV 2019. LNCS, vol. 11561, pp. 443–452. Springer, Cham (2019). https://doi.org/10.1007/978-3-030-25540-4_26
17. LeCun, Y., Cortes, C., Burges, C.: MNIST handwritten digit database (2010)
18. Lecuyer, M., Atlidakis, V., Geambasu, R., Hsu, D., Jana, S.: Certified robustness to adversarial examples with differential privacy. In: 2019 IEEE Symposium on Security and Privacy (SP), pp. 656–672. IEEE (2019)
19. Leino, K., Wang, Z., Fredrikson, M.: Globally-robust neural networks. In: International Conference on Machine Learning (ICML) (2021)
20. Madry, A., Makelov, A., Schmidt, L., Tsipras, D., Vladu, A.: Towards deep learning models resistant to adversarial attacks. In: International Conference on Learning Representations (2018)
21. Mirman, M., Gehr, T., Vechev, M.: Differentiable abstract interpretation for provably robust neural networks. In: International Conference on Machine Learning, pp. 3578–3586. PMLR (2018)
22. Pakdamanian, E., Sheng, S., Baee, S., Heo, S., Kraus, S., Feng, L.: Deeptake: prediction of driver takeover behavior using multimodal data. In: Proceedings of the 2021 CHI Conference on Human Factors in Computing Systems. CHI '21, Association for Computing Machinery, New York, NY, USA (2021). https://doi.org/10.1145/3411764.3445563
23. Papernot, N., et al.: Technical report on the cleverhans v2.1.0 adversarial examples library. arXiv preprint arXiv:1610.00768 (2018)
24. Papernot, N., McDaniel, P., Goodfellow, I., Jha, S., Celik, Z.B., Swami, A.: Practical black-box attacks against machine learning. In: Proceedings of the 2017 ACM on Asia Conference on Computer and Communications Security, pp. 506–519 (2017)
25. Qin, Y., Carlini, N., Cottrell, G., Goodfellow, I., Raffel, C.: Imperceptible, robust, and targeted adversarial examples for automatic speech recognition. In: International Conference on Machine Learning, pp. 5231–5240 (2019). http://proceedings.mlr.press/v97/qin19a.html
26. Raghunathan, A., Steinhardt, J., Liang, P.: Certified defenses against adversarial examples. In: International Conference on Learning Representations (2018). https://openreview.net/forum?id=Bys4ob-Rb

27. Salman, H., Yang, G., Li, J., Zhang, P., Zhang, H., Razenshteyn, I., Bubeck, S.: Provably robust deep learning via adversarially trained smoothed classifiers. In: Proceedings of the 33rd International Conference on Neural Information Processing Systems, pp. 11292–11303 (2019)

28. Singh, G., Gehr, T., Püschel, M., Vechev, M.: Robustness certification with refinement. In: International Conference on Learning Representations (2019). https://openreview.net/forum?id=HJgeEh09KQ

29. Singh, G., Gehr, T., Püschel, M., Vechev, M.: An abstract domain for certifying neural networks. Proc. ACM Program. Lang. **3**(POPL), 1–30 (2019)

30. Szegedy, C., et al.: Intriguing properties of neural networks. In: Bengio, Y., LeCun, Y. (eds.) In: 2nd International Conference on Learning Representations, ICLR 2014, Banff, AB, Canada, April 14–16, 2014, Conference Track Proceedings (2014). http://arxiv.org/abs/1312.6199

31. Tjeng, V., Xiao, K.Y., Tedrake, R.: Evaluating robustness of neural networks with mixed integer programming. In: International Conference on Learning Representations (2019). https://openreview.net/forum?id=HyGIdiRqtm

32. Tramer, F., Carlini, N., Brendel, W., Madry, A.: On adaptive attacks to adversarial example defenses. Advances in Neural Information Processing Systems 33 (2020)

33. Weng, L., et al.: Towards fast computation of certified robustness for relu networks. In: International Conference on Machine Learning, pp. 5276–5285. PMLR (2018)

34. Wong, E., Kolter, Z.: Provable defenses against adversarial examples via the convex outer adversarial polytope. In: International Conference on Machine Learning, pp. 5286–5295. PMLR (2018)

35. Yang, G., Duan, T., Hu, J.E., Salman, H., Razenshteyn, I., Li, J.: Randomized smoothing of all shapes and sizes. In: International Conference on Machine Learning, pp. 10693–10705. PMLR (2020)

36. Zhang, H., Yu, Y., Jiao, J., Xing, E., El Ghaoui, L., Jordan, M.: Theoretically principled trade-off between robustness and accuracy. In: International Conference on Machine Learning, pp. 7472–7482. PMLR (2019)

CEG4N: Counter-Example Guided Neural Network Quantization Refinement

João Batista P. Matos Jr.[1](✉) [iD], Iury Bessa[1] [iD], Edoardo Manino[2] [iD],
Xidan Song[2] [iD], and Lucas C. Cordeiro[1,2] [iD]

[1] Federal University of Amazonas, Manaus-AM, Brazil
`jbpmj@icomp.ufam.edu.br`, `iurybessa@ufam.edu.br`
[2] Univeristy of Manchester, Machester, UK
{`eduardo.manino,xidan.song,lucas.cordeiro`}`@manchester.ac.uk`

Abstract. Neural networks are essential components of learning-based software systems. However, deploying neural networks in low-resource domains is challenging because of their high computing, memory, and power requirements. For this reason, neural networks are often quantized before deployment, but existing quantization techniques tend to degrade network accuracy. We propose Counter-Example Guided Neural Network Quantization Refinement (CEG4N). This technique combines search-based quantization and equivalence verification: the former minimizes the computational requirements, while the latter guarantees that the network's output does not change after quantization. We evaluate CEG4N on a diverse set of benchmarks, including large and small networks. Our technique successfully quantizes the networks in our evaluation while producing models with up to 72% better accuracy than state-of-the-art techniques.

Keywords: Robust quantization · Neural network quantization · Neural network equivalence · Counter example guided optimization

1 Introduction

Neural networks (NNs) are becoming essential in many applications such as autonomous driving [6], security, medicine, and business [2]. However, current state-of-the-art NNs often require substantial compute, memory, and power resources, limiting their applicability [9].

In this respect, quantization techniques help reduce the network size and its computational requirements [9,16,24]. Here, we focus on *quantization* techniques, which aim at reducing the number of bits required to represent the neural network weights [16]. A desirable quantization technique produces the smallest neural network possible from the quantization perspective. However, at the same time, quantization affects the functional behavior of the resulting neural network by making them more prone to erratic behavior due to loss of accuracy [18]. For this reason, existing techniques monitor the degradation in the accuracy of the quantized model with statistical measures defined on the training set [16].

O. Isac et al. (Eds.): NSV 2022/FoMLAS 2022, LNCS 13466, pp. 29–45, 2022.
https://doi.org/10.1007/978-3-031-21222-2_3

However, statistical accuracy measures do not capture the network's vulnerability to malicious attacks. Indeed, there may exist some specific inputs for which the network performance degrades significantly [3,19,27]. For this reason, we reformulate the goal of guaranteeing the accuracy of a quantized model under the notion of *equivalence* [11,12,17,20]. This formal property requires that two neural network models both produce the same output for every input, thus ensuring that the two networks are functionally equivalent [28,30].

We are the first to explore the combination of quantization techniques and equivalence checking in the present work. Doing so guarantees that the quantized model is functionally equivalent to the original one. More specifically, our main scientific contributions are the following:

- We model the equivalence quantization problem as an iterative optimization-verification cycle.
- We propose CEG4N, a counter-example guided neural network quantization technique that provides formal guarantees of NN equivalence.
- We evaluate CEG4N on both large (ACAS Xu [23] and MNIST [26]) and small (Iris [13] and Seeds [8]) benchmarks.
- We demonstrate that CEG4N can successfully quantize neural networks and produce models with similar or better accuracy than a baseline state-of-the-art quantization technique (up to 72% better accuracy).

2 Preliminaries

2.1 Neural Network

NNs are non-linear mapping functions $f : \mathcal{I} \subset \mathbb{R}^n \rightarrow \mathcal{O} \subset \mathbb{R}^m$ consisting of a set of L linked layers, organized as a *direct graph*. Each layer l is connected with the directly preceding layer $l - 1$, *i.e.*, the output of the layer $l - 1$ is the input of the layer l. Exceptions are the first and last layers. The first layer is just a placeholder for the input for the NN while the last layer holds the NN function mapping f. A layer l is composed by a matrix of weights $\mathbf{W_l} \in \mathbb{R}^{n \times m}$ and a bias vector $\mathbf{b_l} \in \mathbb{R}^m$.

The output of a layer is computed by performing the combination of an affine transformation, followed by the non-linear transformation on its input $\mathbf{x_l} \in \mathbb{R}^n$ (see Eq. (1)). Formally, we can describe the function $y_l : \mathbb{R}^n \rightarrow \mathbb{R}^m$ that computes the output of a layer l as follows:

$$y_l(\mathbf{x_l}) = \mathbf{W_l} \cdot \mathbf{x_l} + \mathbf{b_l} \tag{1}$$

and the function that computes the activated output of a layer l as follows:

$$y_l^\sigma(\mathbf{x_l}) = \sigma(y_l(\mathbf{x_l})) \tag{2}$$

where $\sigma : \mathbb{R}^m \rightarrow \mathbb{R}^m$ is the *activation function*. In other words, the output l is the result of the activation function σ applied to the dot product between weight and input, plus the bias. The most popular activation functions are: namely, ReLU, sigmoid (Sigm), and the re-scaled version of the latter known as hyperbolic tangent(TanH) [10]. We focus on the *rectified linear unit* activation function $ReLU = \max\{0, \mathbf{y_1}\}$.

Considering the above, let us denote the input of a NN with L layers as $\mathbf{x} \in \mathbb{I}$, and $f(x) \in \mathcal{O}$ as the output; thus, we have that:

$$f(\mathbf{x}) = \sigma\left(y_L(\sigma\left(y_{L-1}(...(\sigma\left(y_1(\mathbf{x})\right)))\right)\right)\right) \tag{3}$$

2.2 Quantization

Quantization is the process of constraining high precision values (*e.g.*, single-precision floating-point values) to a finite range of lower precision values (*e.g.*, a discrete set such as the integers) [1,16]. The quantization quality is usually determined by a scalar n (the available number of bits) that defines the lower and upper bounds of the finite range. Let us define quantization as a mapping function $\mathcal{Q}_n : \mathbb{R}^{m \times p} \to \mathbb{I}^{m \times p}$, formulated as follow:

$$Q\left(n, A\right) = clip\left(\left\lfloor \frac{A}{q(A, n)} \right\rceil, -2^{n-1}, 2^{n-1} - 1\right) \tag{4}$$

where $A \in \mathbb{R}^{m \times p}$ denotes the continuous value– notice that A can be a single scalar, a vector, or a matrix; n denotes the number of bits for the quantization, $q(A, n)$ denotes a function that calculates the scaling factor for A in respect to a number of bits n, and $\lfloor \cdot \rceil$ denotes rounding to the nearest integer. Defining the scaling factor (see Eq. 5) is an important aspect of uniform quantization [22,25].

The scaling factor is essentially what divides a given range of real values A into an arbitrary number of partitions. Thus, let us define a scaling factor function by $q_n(A)$, a number of bits (bit-width) to be used for the quantization by n, a clipping range by $[\alpha, \beta]$, the scaling factor can be defined as follow:

$$q(A, n) = \frac{\beta - \alpha}{2^n - 1} \tag{5}$$

The min/max of the signal are often used to determining the clipping range values, *i.e.*, $\alpha = \min A$ and $\beta = \max A$. But as we are using symmetric quantization, the clipping values are defined as $\alpha = \beta = \max([|\min A|, |\max A|])$. In practice, the quantization process can produce an integer value that lies outside the range of $[\alpha, \beta]$. To prevent that, the quantization process will have an additional clip step.

Equation (6) shows the corresponding de-quantization function, which computes back the original floating-point value. However, we should note that the de-quantization approximates the original floating-point value.

$$\hat{A} = q(A, 2)Q\left(n, A\right) \tag{6}$$

2.3 NNQuantization

In this section, we discuss how a convolutional or fully-connected NN layer can be quantized in the symmetric mode. Considering l to be any given layer in a NN, let us denote $\mathbf{x}_l$, $\mathbf{W}_l$, and $\mathbf{b}_l$ as the original floating-point input vector, the original floating-point weight matrix, and the original floating-point bias vector, respectively, of the layer l. And applying the de-quantization function from Eq. (6), where, we assume that $A = \hat{A}$. Borrowing from notations used in Sects. 2.1 and 2.2. We can formalize the quantization of a NN layer l as follows:

$$
\begin{aligned}
y_l(\mathbf{x}_l) &= \mathbf{W}_l \cdot \mathbf{x}_l + \mathbf{b}_l \\
&\approx q(\mathbf{W}_l, n_l) Q(n_l, \mathbf{W}_l) \cdot \mathbf{x}_l + q(\mathbf{b}_l, n_l) Q(n_l, \mathbf{b}_l)
\end{aligned}
\tag{7}
$$

Notice that the bias does not need to be re-scaled to match the scale of the dot product. Since we consider maximum scaling factor between $q(\mathbf{W}_l, n_l)$ and $q(\mathbf{b}_l, n_l)$), both the weight and the bias share the same scaling factor in Eq. (7). With that in mind, the formalization of a NN f in Eq. (3) can be reused to formalize a quantized NN as well.

2.4 NNEquivalence

Let $\mathcal{F}$ and $\mathcal{T}$ be two arbitrary NNs, and let $\mathcal{I} \in \mathbb{R}^n$ be the common input space of the two NNs and $\mathcal{O} \in \mathbb{R}^m$ be their common output space. Thus, NN equivalence verification is the problem of proving that $\mathcal{F}$ and $\mathcal{T}$, or more specifically, their corresponding mathematical functions $f : \mathcal{I} \to \mathcal{O}$, $t : \mathcal{I} \to \mathcal{O}$ are equivalent. In essence, by proving the equivalence between two neural networks, one can prove that both NNs produce the same outputs for the same set of inputs. Currently, the literature reports the following definition of equivalence.

Definition 1 (Top-1-Equivalence [7,30]). *Two NNs f and t are Top-1-equivalent, if $\arg\max \ f(x) = \arg\max \ t(x)$, for all $x \in \mathcal{I}$.*

Let us formalise the notion of *Top-1 Equivalence* in first-order logic. This is necessary for the comprehension of the equivalence verification explained in the following sections of the paper. But first, we formalize some essential assumptions for the correctness of the equivalence properties.

Assumption 1. *Let $f(x)$ be the output of the NN $\mathcal{F}$ in real arithmetic (without quantization). It is assumed that $\arg\max f(x) = y$ such that $x \in \mathcal{H}$.*

Assumption 2. *Let $f^q(x)$ be the output of the NN $\mathcal{F}$ in a quantized form. There is set of numbers of bits $\mathcal{N}$ such that $\arg\max f(x) = \arg\max f^q(x) = y$ for all $x \in \mathcal{H}$.*

Note that the quantization of the NN f that results in the NN $f^q(x)$ depends on the number of bits N. Refer to Eq. (7) to understand the relationship between N and f^q.

An instance of a equivalence verification is given by a conjunction of constraints on the input $\psi_x(x)$, the output $\psi_y(y)$ and the NNs f and f^q. $\psi(f, f^q, x, y) = \psi_x(x) \rightarrow \psi_y(y)$. We denote $\psi_y(y)$ the equivalence constraint. Let $\bar{x} = x + \hat{x}$ such that $|x + \hat{x}|_\infty \leq \epsilon$, consider $\bar{x} \in \mathcal{H}$ and $y \in \mathcal{G}$. Taking from Definition 1, we have that:

- $\psi_x(x)$ is an equivalence property such that $\psi_x(x) \leftrightarrow \bar{x} \in \mathcal{H}$
- $\psi_y(y)$ is an equivalence property such that $\psi_y(y) \leftrightarrow \arg\max f^q(x) = y$

Note that, to prove the equivalence of f and f^q, one may prove that the property $\psi(f, f^q, x, y)$ holds for any given x and y. This approach may not be feasible. But proving that $\psi(f, f^q, x, y)$ does not hold for some x and y is a more tractable approach. If we do so, we can provide a counter-example.

2.5   Verification of NNProperties

In this paper, we use the classic paradigm of SMT verification. In this paradigm, the property to check (e.g., equivalence) and the computational model (e.g., the neural networks) are encoded as a first-order logic formula, which is then checked for satisfiability. Moreover, to keep the problem decidable, SMT restricts the full expressive power of first-order logic to a decidable fragment.

SMT formulas can capture the complex relationship between variables, holding real, integer values and other data types. If it is possible to assign values to such variables that a formula is evaluated as true, then the formula is said to be *satisfiable*. On the other hand, if it's not possible to assign such values, the formula is said to be *unsatisfiable*.

Given a NN $\mathcal{F}$ and its mathematical function f, a set of safe input instances $\mathcal{H} \in \mathbb{R}^n$, and a safe domain $\mathcal{G} \subseteq \mathcal{O}^m$ – both defined as a set of constraints, safety verification is concerned with the question of whether there exist an instance $x \in \mathcal{H}$ such that $f(x) \notin \mathcal{G}$. An instance of a safety verification is given by a conjunction of constraints on the input $\psi_x(x)$, the output $\psi_y(y)$ and the NN f. $\psi(f, x, y) = \psi_x(x) \rightarrow \psi_y(y)$ is said to be satisfiable if there exists some $x \in \mathcal{H}$ such that $f(x)$ returns y for the input x and $\psi(f, x, y)$ does not hold.

3   Counter-Example Guided Neural Network Quantization Refinement (CEG4N)

We define *robust quantization (RQ)* to describe the problem of maximizing the quantization of a NN while keeping the equivalence between the original model and the quantized one (see Definition 2). Borrowing from the notations used in Sect. 2, we formally define RC as follows.

Definition 2 (Robust Quantization). *Let f be the reference NN and $\mathcal{H} \in \mathbb{R}^n$ be a set of inputs instances. We define robust quantization as a process that performs the quantization of f hence resulting in a quantized model f^q such that $\arg\max f(x) \iff \arg\max f^q(x) \; \forall \; x \in \mathcal{H}$.*

From the definition discussed in Sect. 2.4, we preserve the equivalence between the mathematical functions f and f^q associated with the NNs. In the RC, we shift the focus from the original NN to the quantized NN, *i.e.*, we assume that f is safe (or robust) and use it as a reference to define the safety properties we expect for f^q. By checking the equivalence of f and f^q, we can state that f^q is robust, and therefore, we achieve a *robust quantization*. In more details, consider a NN f with L layers. The quantization of f assumes there is a set $\mathcal{N} = \{n_1, n_2, \cdots, n_L\}$, where $n_l \in \mathcal{N}$ represents the number of bits that should be used to quantize the l-th layer in f. In our robust quantization problem, we obtain a sequence $\mathcal{N}$ for which each $n \in \mathcal{N}$ is minimized (e.g., one could minimize the sum of all $n \in \mathcal{N}$) and the equality between f and f^q is satisfied.

3.1   Robust Quantization as a Minimization Problem

We consider the robust quantization of a NN as an iterative minimization problem. Each iteration is composed of two complementary sub-problems. First, we need to minimize the quantization bit widths, that is, finding a candidate set $\mathcal{N}$. Second, we need to verify the equivalence property, that is, checking if a NN quantized with the bit widths in $\mathcal{N}$ is equivalent to the original NN. If the latter fails, we iteratively return to the minimization sub-problem with additional information. More specifically, we formalize the first optimization sub-problem as follows.

Optimization sub-problem o:

$$\textbf{Objective:} \quad \mathcal{N}^o = \arg\min_{n_1^o, \ldots, n_L^o} \sum_{l \in \mathbb{N}_{l \leq L}} n_l$$

$$\textbf{s.t:} \quad \arg\max f(x) = \arg\max f^q(x), \; \forall \; x \in \mathcal{H}_{\text{CE}}^o \qquad (8)$$

$$n_l \geq \underline{N} \; \forall \; n_l \in \mathcal{N}^o$$

$$n_l \leq \overline{N} \; \forall \; n_l \in \mathcal{N}^o$$

where f is the mathematical function associated with the NN $\mathcal{F}$ and f^q is the quantized mathematical function associated with the NN $\mathcal{F}$, $\mathcal{H}_{\text{CE}}^o$ is a set of counter-examples available at iteration o. Consider $\underline{N}$ and $\overline{N}$ as the minimum and the maximum bit width allowed to be used in the quantization; these parameters are constant. $\overline{N}$ ensures two things, it gives an upper bound to the quantization bit width, and provides a termination criteria, if a candidate $\mathcal{N}^o$ such that $n_l = \overline{N}$ for every $n_l \in \mathcal{N}^o$, the optimization is stopped because it reached our **Assumption** 2. In particular, our **Assumption** 2 ensures the termination of CEG4N, and it is build over the fact that there is a set of $\overline{N}$ for which the quantization introduces a minimal amount of error to NN. In any case, if

CEG4N proposes a quantization solution equal to the $\overline{N}$, this solution is verified as well, and in case the verification returns a counter-example, CEG4N finishes with failure. Finally, note that $\mathcal{H}_{CE}^o$ is an iterative parameter, meaning its value is updated at each iteration o. This is done based on the verification sub-problem (formalized below).

Verification sub-problem o:
In the verification sub-problem o, we check whether the $\mathcal{N}^o$ generated by the optimization sub-problem o satisfies the following equivalence property:

$$\psi(f, f^q, x, y) = \psi_x(x) \rightarrow \psi_y(y)$$

if $\psi_x(x) \rightarrow \psi_y(y)$ holds for the candidate $\mathcal{N}^o$, the optimization halts and $\mathcal{N}^o$ is declared as solution; otherwise, a new counter-example x_{CE} is generated. Iteration $o+1$ starts where iteration o stopped. That is, the optimization sub-problem $o + 1$ receives as parameter a set of $\mathcal{H}_{CE}^{o+1}$ such that $\mathcal{H}_{CE}^{o+1} = \mathcal{H}_{CE}^o \cup x_{CE}$.

3.2   The CEG4NFramework Implementation

We propose CEG4N framework, which is a counterexample-guided optimization approach to solve the robust quantization problem. In this approach, we consider combining two main modules to solve the two sub-problems presented in Sect. 3.1: the optimization of the bit widths for the quantization and the verification of the NN equivalence. The first module that solves the optimal bit width problem roughly takes in a NN and generates quantized NN candidates. Then, the second module takes in the candidates and verifies their equivalence to the original model.

Figure 1 illustrates the overall architecture of the CEG4N framework. It also shows how each framework's module interacts with the other and in what sequence. The *GA module* is an instance of a Genetic Algorithm. The GA module expects two main parameters, NN and a set of counter-examples $\mathcal{H}_{CE}$ We can also specify a maximum number of generations the algorithm is allowed to run and lower and upper bounds to restrict the possible number of bits. Once the GA module produces a candidate, that is, a sequence of bit widths, for each layer of the neural network, CEG4N generates the C-Abstraction code for the original model and the quantized candidate and then checks their equivalence. Each check for this equivalence property is exported to a unique verification test case. Then, it triggers the execution of the verifier for each verification test case and awaits the verifier output. Here, *Verifier module* is an instance of a formal verifier (i.e., a Bounded Model Checker (BMC), namely, ESBMC [15]). This step is done sequentially, meaning each verification is run once the last verification terminates.

Once all verification test cases terminate, CEG4N collect and process all outputs and checks whether any counter-example has been found. If so, it updates the set of counter-examples $\mathcal{H}_{CE}'$ and triggers the GA module execution again, thus initiating a new iteration of CEG4N. If no counter-example is found,

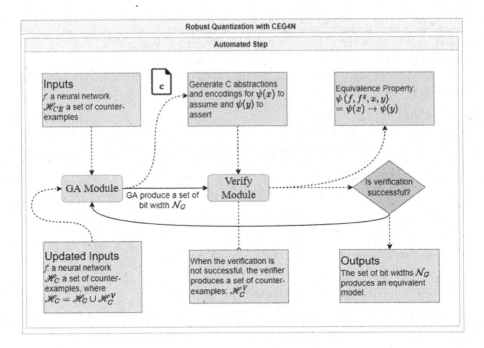

Fig. 1. CEG4N architecture overview, highlighting the relationship between the main modules, and their inputs and outputs.

CEG4N considers the verification successful and terminates the quantization process outputting the found solution.

We work with two functional versions of the NN. The GA module works with a functional NN written in Python, while the verifier module works with a functional version of the NN written in C. The two models are equivalent since they share the same parameters; the python model loads the parameters to a framework built over Pytorch [29]. The C version loads the weights into a framework designed and developed in C to work correctly with the verifier idioms and annotations. We provide more details regarding the C implementations of the NNs in Sect. A.2.

4   Experimental Evaluation

This section describes our experimental setup and benchmarks, defines our objectives, and presents the results.

4.1   Description of the Benchmarks

We evaluate our methodology on a set of feedforward NN classification models extracted from the literature [10,23,26]. We chose these specific ones based on their popularity in previous NN robustness and equivalence verification studies

[10, 30]. Additionally, we include a few other NN models to cover a broader range of NN architectures (e.g., NN size, number of neurons).

ACAS Xu. The airborne collision avoidance system for unmanned aircraft ACAS Xu dataset [23] is derived from 8 specifications (features boundaries and expected outputs). ACAS Xu features are sensor data indicating the speed, present course of the aircraft, and the position and speed of any nearby intruder aircraft. An ACAS Xu NN is expected to give appropriate navigation advisories for a given input sensor data. The expected outputs indicate that either the aircraft is clear-of-conflict, or it should take soft or hard turns to avoid the collision. We evaluated CEG4N on 5 pre-trained NNs, each containing 8 layers and 300 ReLU nodes each. The pre-trained NNs were obtained from the VNN-COMP2021 [5] benchmarks[1]

MNIST. MNIST is a popular dataset [26] for image classification. The dataset contains 70,000 gray-scale images with uniform size of 28×28 pixels, where the original pixel values from the integer range $[0, 255]$ are rescaled to the floating-point range $[0, 1]$. We evaluated CEG4N on two NNs with 2 layers, one with 10 ReLU nodes each and another with 25 and 10 ReLU nodes. The NNs followed the architecture of models described by the work of Eleftheriadis et al. [10].

Seeds. The Seeds dataset [8] consists of 210 samples of wheat grain belonging to three different species, namely Kama, Rosa and Canadian. The input features are seven measurements of the wheat kernel geometry scaled between [0,1]. We evaluated CEG4N on 2 NNs, containing 1 layer, one containing 15 ReLU nodes, and the other containing 2 ReLU nodes. Both NNs were trained for the CEG4N evaluation.

Iris. The Iris flower dataset [13] consists of 50 samples from three species of Iris flower (*Iris setosa, Iris virginica and Iris versicolor*). The dataset is a popular benchmark in machine learning for classification, and the data is composed of records of real value measurements of the width and length of sepals and petals of the flowers. The data was scaled to [0,1]. We evaluated CEG4N on 2 NNs, one of them containing 2 layers with 20 ReLU nodes and the other having only one layer with 3 ReLU nodes. Both NNs were trained for the CEG4N evaluation.

4.2   Setup

Genetic Algorithm. As explained in Sect. 3.1, we quantize the NNs with a NSGA-II Genetic Algorithm module. We set the upper and lower bounds for the allowed bit widths to 2 and 52 in all experiments. The lower bound was

---

[1] The pre-trained weight for the ACAS Xu benchmarks can be found in the following repository: https://github.com/stanleybak/vnncomp2021.

chosen because 2 is the first valid integer that does not break our quantization formulas. The upper bound was chosen to match the significand of the double-precision float format IEEE 754-1985 [21]. The upper bound value could be higher depending on the precision of weights parameters of the NN, as the scaling factor could lead the quantization to large integer values. However, as we wanted the framework to work on every NN in our experimentation setup without further steps, we restricted the clipping range to a comfortable number to avoid integer overflow.

Furthermore, we allow the GA to run for 110 generations for each layer in the NN. This number of generations was defined after extensive preliminary tests, which confirmed that GA could reach the optimal solution in most cases (see Table 3 in Appendix A.4). Lastly, we randomly select the initial set of counter-examples $\mathcal{H}$ from the benchmark set of each case study. The samples in $\mathcal{H}$ do not necessarily have to be *counter-examples*, and any valid concrete input can be specified. Our choice is justified by the practical aspect of using samples from the benchmark set.

Equivalence Properties. One input sample was selected for each output class and used to define the equivalence properties. Due to the high dimensional number of the features in the MNIST study case, we proposed a different approach when specifying the equivalence properties for the equivalence verification. We considered three different approaches: 1) one in which we considered all features in the input domain; 2) another one in which we considered only a subset of 10 out of the 784 features in the input domain; 3) a last one in which we considered only a subset of 4 out of the 784 features in the input domain. The subset of features in cases 2 and 3 was randomly selected.

Availability of Data and Tools. Our experiments are based on a set of publicly available benchmarks. All tools, benchmarks, and results of our evaluation are available on a supplementary web page https://zenodo.org/record/6791964.

4.3   Objectives

Considering the benchmarks given in Sect. 4.1, our evaluation has the following two experimental goals:

EG1 (**robustness**) Show that the CEG4N framework can generate robust quantized NNs.

EG2 (**accuracy**) Show that the quantized NNs do not have a significant drop in accuracy compared to other quantization techniques.

4.4   Results

In our first set of experiments, we want to achieve our first experimental goal **EG1**. We want to show that our technique CEG4N can successfully generate quantized NNs that are verifiably equivalent to the original NNs. As a secondary goal, we want to perform an empirical scalability study to help us evaluate the computational demands for quantizing and verifying the equivalence of NNs models. Our findings are summarized in Table 1.

Table 1. Summary of the CEG4N executions, including the models, number of features, the number of bits per layer, and the status.

Model	Features	Equivalence Properties	Iterations	Bits	Status
iris_3	4	3	1	4, 3	*completed*
seeds_2	7	3	1	4, 3	*completed*
seeds_15	7	3	1	4, 2	*completed*
acasxu_1	5	6	1	6, 8, 7, 7, 9, 7, 6	*completed*
acasxu_2	5	7	1	10, 9, 9, 9, 7, 7, 10	*completed*
acasxu_3	5	7	1	5, 9, 10, 7, 8, 8, 5	*completed*
acasxu_4	5	7	1	8, 9, 14, 9, 10, 10, 7	*completed*
acasxu_5	5	7	1	6, 12, 8, 8, 10, 10, 10	*completed*
mnist_10	5	10	1	4, 3	*completed*
	10	10	1	4, 3	*completed*
	784	10	0	4, 3	*timeout*
mnist_25	5	10	1	3, 3	*completed*
	10	10	1	3, 3	*completed*
	784	10	0	3, 3	*timeout*

All the CEG4N runs that were completed successfully took only 1 iteration to find a solution. However, we observed that four of the CEG4N attempts to find a solution for MNIST models resulted in a timeout. We attribute this observation to a mix of factors. First is the high number of features in the MNIST problem. Second, the network's overall architecture requires many arithmetic operations to compute the model's output. Finally, we also observed that it took only a few minutes for CEG4Nto find a solution to the Iris, Seeds, and Acas Xu benchmarks. In contrast, on MNIST, it took hours to either find a solution or fail with a timeout.

> These results answer our **EG1**: overall, these experiments show that CEG4N can successfully produce robust quantized models. Although, one should notice that for larger NNs models, scalability should be a point of concern due to our verifier stage.

In our second set of experiments, we want to achieve our second experimental goal **EG2**. We primarily want to understand the impact of the quantization performed by CEG4N on the accuracy of the NNs compared to other quantization techniques. Due to our research's novelty, no existing techniques lend themselves to a fair comparison. For this reason, we take a recent post-training quantization technique called GPFQ [31] and modify it to our needs. GPFQ [31] is a greedy path-following quantization technique that also produces quantized models with floating/double-precision values. It works by iterating over each layer of the NN and quantizing each neuron sequentially. More specifically, a greedy algorithm minimizes the error between the original neural output and the quantized neuron.

Table 2 summarizes the accuracy of the models quantized using CEG4N and GPFQ. Note that we do not report the accuracy of the Acas Xu models because the original training and test datasets are not public.

Table 2. Comparison of Top-1 accuracy for NNs quantized using CEG4N and GPFQ

Model	Method	Ref Acc (%)	Quant Acc (%)	Acc Drop (%)
iris_3	CEG4N	93.33	83.33	10.0
	GPFQ		23.33	70.0
seeds_2	CEG4N	88.09	85.71	2.38
	GPFQ		64.28	23.81
seeds_15	CEG4N	90.04	85.71	4.33
	GPFQ		40.47	49.57
mnist_10	CEG4N	91.98	86.7	5.28
	GPFQ		91.29	0.69
mnist_25	CEG4N	93.68	92.57	1.11
	GPFQ		92.59	1.09

Our findings show that the highest drops in accuracy happen on the Iris benchmark (10% for CEG4N and 70% drop for GPFQ). In contrast, the lowest drops in accuracy happen on mnist_25 for CEG4N and on mnist_10 for GPFQ. Overall, the accuracy of models quantized with CEG4N are better on the Iris and Seeds benchmarks, while the accuracy of models quantized with GPFQ are better on the mnist benchmarks, but only by a small margin. Our understanding is that GPFQ shows high drops in accuracy for smaller NNs because the number of neurons in each layer is small. As GPFQ focuses on each neuron individually, it may not be able to find a good global quantization.

> These results answer our **EG2**: overall, these experiments show that CEG4N can successfully produce quantized models with superior or similar accuracy to other state-of-the-art techniques.

4.5 Limitations

Although we showed in our evaluation that the CEG4N framework can generate a quantized neural network while keeping the equivalence between the original NN and the quantized NN, we note that the architecture of the NN used in the evaluation does not fully reflect state-of-the-art NN architectures. The NNs used in our evaluation have few layers and only hundreds of ReLU nodes, while state-of-the-art NNs may have hundreds of layers and thousands of ReLU nodes. The main bottleneck is state-of-the-art verification algorithms, which currently do not scale to large neural networks. As it is, our technique could only quantized 80% of the NN in our experimental evaluation.

In addition, the field of research on NN equivalence is relatively new and there is no well-established set of benchmarks that works in this field could benefit from [10]. Furthermore, our work is the first to propose a framework that mixes NN quantization and NN equivalence verification. There is no comparable methodology in the literature we could compare our approach with.

5 Conclusion

We presented a new method for NN quantization, called CEG4N, a post-training NN quantization technique that provides formal guarantees of NN equivalence. This approach leverages a counter-example guided optimization technique, where an optimization-based quantizer produces quantized model candidates. A state-of-the-art C verifier then checks these candidates to prove the equivalence of the quantized candidates and the original models or refute that equivalence by providing a counter-example. This counter-example is then passed back to the quantized to guide it to search for a feasible candidate.

We evaluate the CEG4N method on four benchmarks, including large models (ACAS Xu and MNIST) and smaller models (Iris and Seeds). We successfully demonstrate the application of the CEG4N for NN quantization, where it could successfully quantize the networks while producing models with up to 72% better accuracy than state-of-the-art techniques. However, CEG4N can only handle a restricted set of NNs models, and further work needs to scale the CEG4N applicability on a broader set of NNs models (e.g., NNs models with a more significant number of layers and neurons and higher numbers of input features).

For future work, we could explore other quantization techniques, which are not limited to search-based quantization and other promising equivalence verification techniques using a MILP approach [30] or an SMT-based approach [10]. Combining different quantization and equivalence verification techniques can enable CEG4N to achieve better scalability and quantization rates. Another interesting future work relates to the possibility of mixing quantization approaches that generate quantized models, which operate entirely on integer arithmetic; this can potentially improve the verification step scalability of the CEG4N.

Acknowledgment. The work is partially funded by EPSRC grant EP/T026995/1 entitled "EnnCore: End-to-End Conceptual Guarding of Neural Architectures" under *Security for all in an AI-enabled society.*

A   Appendices

A.1   Implementation of NNs in Python.

The NNs were built and trained using the Pytorch library [29]. Weights of the trained models were then exported to the ONNX [4] format, which can be interpreted by Pytorch and used to run predictions without any compromise in the NNs performance.

A.2   Implementation of NNs abstract models in C.

In the present work, we use the C language to implement the abstract representation of the NNs. It allows us to explicitly model the NN operations in their original and quantized forms and apply existing software verification tools (e.g., ESBMC [14]). The operational C-abstraction models perform double-precision arithmetic. Although, we must notice that the original and quantized only diverge on the precision of the weight and bias vectors that are embedded in the abstractions code.

A.3   Encoding of Equivalence Properties

Suppose, a NN F, for which $x \in \mathcal{H}$ is a safe input and $y \in \mathcal{G}$ is the expected output of f the input. We now show how one can specify the equivalence properties. For this example, consider that the function f can produce the outputs of F in floating-point arithmetic, while fq produces the outputs of F in fixed-point arithmetic (*i.e.* quantization). First, the concrete NN input x is replaced by a non-deterministic one, which is achieved using the command **nondet_float** from the ESBMC.

Listing 1.1. Definition of concrete and symbolic input domain in *EBMC.*

```
float  x0 = −1.0;
float  x1 = 1.0;
float  s0 = nondet_float ();
float  s1 = nondet_float ();
```

Listing 1.2. Definition of input constraints in *EBMC.*

```
const float EPS = 0.5;
__ESBMC_assume ( x0 − EPS <= s0 && s0 <= x0 + EPS );
__ESBMC_assume ( x1 − EPS <= s1 && s1 <= x1 + EPS );
```

Listing 1.3. Definition of output constraints in *EBMC.*

```
__ESBMC_assert ( f ( s0 , s1 ) == fq ( s0 , s1 ));
```

A.4   Genetic Algorithm Parameters Definition

In Table 3, we report a summary of experiments conducted to tune the parameters of the Genetic Algorithm, more precisely, the number of generations. For example, a NN with 2 layers would require a brute force algorithm to search for 52^2 combinations of bits widths for the quantization. Similarly, a NN with 7 layers would require a brute force algorithm to search for 52^7 combinations of bits widths. We conducted a set of experiments where we ran the GA one hundred times with a different number of generations options ranging from 50 to 1000. In addition, we fixed the population size to 5. From our findings, the GA needs about 100 to 110 generations per layer to find the optimal bit width solution for each run.

Table 3. Summary of experiments for tuning Genetic Algorithm Parameters.

Number of layers	Generations	Population	Percentage of optimal solutions
7	800	5	100
7	750	5	100
7	700	5	98
7	50	5	0
2	250	5	100
2	200	5	100
2	150	5	96
2	50	5	30

References

1. Abate, A., et al.: Sound and automated synthesis of digital stabilizing controllers for continuous plants. In: Frehse, G., Mitra, S. (eds.) Proceedings of the 20th International Conference on Hybrid Systems: Computation and Control, HSCC 2017, Pittsburgh, PA, USA, 18–20 April 2017, pp. 197–206. ACM (2017). https://doi.org/10.1145/3049797.3049802
2. Abiodun, O.I., Jantan, A., Omolara, A.E., Dada, K.V., Mohamed, N.A., Arshad, H.: State-of-the-art in artificial neural network applications: a survey. Heliyon. **4**(11), e00938 (2018). https://doi.org/10.1016/j.heliyon.2018.e00938, https://www.sciencedirect.com/science/article/pii/S2405844018332067
3. Albarghouthi, A.: Introduction to neural network verification. arXiv:2109.10317 (2021)
4. Bai, J., Lu, F., Zhang, K., et al.: ONNX: Open neural network exchange (2019). https://github.com/onnx/onnx
5. Bak, S., Liu, C., Johnson, T.: The second international verification of neural networks competition (VNN-COMP 2021): Summary and results (2021)

6. Bojarski, M., et al.: End to end learning for self-driving cars. arXiv:1604.07316 (2016)
7. Kleine Büning, M., Kern, P., Sinz, C.: Verifying equivalence properties of neural networks with ReLU activation functions. In: Simonis, H. (ed.) CP 2020. LNCS, vol. 12333, pp. 868–884. Springer, Cham (2020). https://doi.org/10.1007/978-3-030-58475-7_50
8. Charytanowicz, M., Niewczas, J., Kulczycki, P., Kowalski, P.A., Łukasik, S., Żak, S.: Complete gradient clustering algorithm for features analysis of x-ray images. In: Piętka, E., Kawa, J. (eds) Information Technologies in Biomedicine. AINSC, vol. 69, pp. 15–24. Springer, Heidelberg (2010). https://doi.org/10.1007/978-3-642-13105-9_2
9. Cheng, Y., Wang, D., Zhou, P., Zhang, T.: A survey of model compression and acceleration for deep neural networks. arXiv:1710.09282 (2017)
10. Eleftheriadis, C., Kekatos, N., Katsaros, P., Tripakis, S.: On neural network equivalence checking using SMT solvers. arXiv:2203.11629 (2022)
11. Esser, S.K., Appuswamy, R., Merolla, P., Arthur, J.V., Modha, D.S.: Backpropagation for energy-efficient neuromorphic computing. In: Cortes, C., Lawrence, N., Lee, D., Sugiyama, M., Garnett, R. (eds.) Advances in Neural Information Processing Systems, vol. 28. Curran Associates, Inc. (2015). https://proceedings.neurips.cc/paper/2015/file/10a5ab2db37feedfdeaab192ead4ac0e-Paper.pdf
12. Farabet, C., et al.: Large-scale FPGA-based convolutional networks (2011)
13. Fisher, R.A.: The use of multiple measurements in taxonomic problems. Ann. Eugen. **7**, 179–188 (1936)
14. Gadelha, M.R., Menezes, R.S., Cordeiro, L.C.: ESBMC 6.1: automated test case generation using bounded model checking. Int. J. Softw. Tools Technol. Transf. **23**(6), 857–861 (2020). https://doi.org/10.1007/s10009-020-00571-2
15. Gadelha, M.R., Monteiro, F.R., Morse, J., Cordeiro, L.C., Fischer, B., Nicole, D.A.: ESBMC 5.0: an industrial-strength C model checker. In: 2018 33rd IEEE/ACM International Conference on Automated Software Engineering (ASE), pp. 888–891 (2018). https://doi.org/10.1145/3238147.3240481
16. Gholami, A., Kim, S., Dong, Z., Yao, Z., Mahoney, M.W., Keutzer, K.: A survey of quantization methods for efficient neural network inference. arXiv:2103.13630 (2022)
17. Han, S., Pool, J., Tran, J., Dally, W.J.: Learning both weights and connections for efficient neural network. arXiv:1506.02626 (2015)
18. Hooker, S., Courville, A.C., Dauphin, Y., Frome, A.: Selective brain damage: measuring the disparate impact of model pruning. arXiv:1911.05248 (2019)
19. Huang, X., et al.: Safety and trustworthiness of deep neural networks: a survey. arXiv:1812.08342 (2018)
20. Hubara, I., Courbariaux, M., Soudry, D., El-Yaniv, R., Bengio, Y.: Quantized neural networks: training neural networks with low precision weights and activations. arXiv:1609.07061 (2017)
21. IEEE: IEEE standard for floating-point arithmetic. IEEE Std. 754–2019 (Revision of IEEE 754–2008), pp. 1–84 (2019). https://doi.org/10.1109/IEEESTD.2019.8766229
22. Jacob, B., et al.: Quantization and training of neural networks for efficient integer-arithmetic-only inference. CoRR abs/1712.05877 (2017). arxiv:1712.05877
23. Julian, K.D., Lopez, J., Brush, J.S., Owen, M.P., Kochenderfer, M.J.: Policy compression for aircraft collision avoidance systems. In: 2016 IEEE/AIAA 35th Digital Avionics Systems Conference (DASC), pp. 1–10 (2016). https://doi.org/10.1109/DASC.2016.7778091

24. Kirchhoffer, H., et al.: Overview of the neural network compression and representation (NNR) standard. IEEE Trans. Circ. Syst. Video Technol. 1 (2021). https://doi.org/10.1109/TCSVT.2021.3095970
25. Krishnamoorthi, R.: Quantizing deep convolutional networks for efficient inference: a whitepaper. CoRR abs/1806.08342 (2018). arxiv:1806.08342
26. LeCun, Y., Cortes, C.: The MNIST database of handwritten digits (2005)
27. Liu, C., Arnon, T., Lazarus, C., Barrett, C.W., Kochenderfer, M.J.: Algorithms for verifying deep neural networks. Found. Trends Optim. **4**, 244–404 (2021)
28. Narodytska, N., Kasiviswanathan, S.P., Ryzhyk, L., Sagiv, S., Walsh, T.: Verifying properties of binarized deep neural networks. In: AAAI (2018)
29. Paszke, A., et al.: Pytorch: an imperative style, high-performance deep learning library. In: Wallach, H., Larochelle, H., Beygelzimer, A., d'Alché-Buc, F., Fox, E., Garnett, R. (eds.) Advances in Neural Information Processing Systems, vol. 32, pp. 8024–8035. Curran Associates, Inc. (2019). http://papers.neurips.cc/paper/9015-pytorch-an-imperative-style-high-performance-deep-learning-library.pdf
30. Teuber, S., Buning, M.K., Kern, P., Sinz, C.: Geometric path enumeration for equivalence verification of neural networks. In: 2021 IEEE 33rd International Conference on Tools with Artificial Intelligence (ICTAI), November 2021. https://doi.org/10.1109/ictai52525.2021.00035, http://dx.doi.org/10.1109/ICTAI52525.2021.00035
31. Zhang, J., Zhou, Y., Saab, R.: Post-training quantization for neural networks with provable guarantees. arXiv preprint arXiv:2201.11113 (2022)

Minimal Multi-Layer Modifications of Deep Neural Networks

Idan Refaeli and Guy Katz[✉]

The Hebrew University of Jerusalem, Jerusalem, Israel
{idan.refaeli,g.katz}@mail.huji.ac.il

Abstract. Deep neural networks (DNNs) have become increasingly popular in recent years. However, despite their many successes, DNNs may also err and produce incorrect and potentially fatal outputs in safety-critical settings, such as autonomous driving, medical diagnosis, and airborne collision avoidance systems. Much work has been put into detecting such erroneous behavior in DNNs, e.g., via testing or verification, but removing these errors after their detection has received lesser attention. We present here a new tool, called 3M-DNN, for *repairing* a given DNN, which is known to err on some set of inputs. The novel repair procedure implemented in 3M-DNN computes a modification to the network's weights that corrects its behavior, and attempts to minimize this change via a sequence of calls to a backend, black-box DNN verification engine. To the best of our knowledge, our method is the first one that allows repairing the network by simultaneously modifying multiple layers. This is achieved by splitting the network into sub-networks, and applying a single-layer repairing technique to each component. We evaluated 3M-DNN tool on an extensive set of benchmarks, obtaining promising results.

1 Introduction

The popularity of *deep neural networks (DNNs)* [21] has increased significantly over the past few years. DNNs are machine-learned artifacts, trained using a finite training set of examples; and they are capable of correctly handling previously-unseen inputs. DNNs have shown great success in many application domains, such as image recognition [10,39], audio transcription [50], language translation [52], and even in safety-critical domains such as medical diagnosis [38], autonomous driving [6], and airborne collision avoidance [28].

Despite their evident success, DNNs can sometimes contain bugs. This has been demonstrated repeatedly: in one famous example, Goodfellow et al. [22] showed that slight perturbations to a DNN's input could lead to misclassification—a phenomenon now known as susceptibility to *adversarial perturbations*. In another case, Liu et al. [44] showed how DNNs are vulnerable to Trojan attacks. These issues, and others, combined with the increasing integration of DNNs into safety-critical systems, have created a surge of interest in

O. Isac et al. (Eds.): NSV 2022/FoMLAS 2022, LNCS 13466, pp. 46–66, 2022.
https://doi.org/10.1007/978-3-031-21222-2_4

establishing their correctness. A great deal of effort has been put into developing methods for testing DNNs [57], and, more recently, also into verifying them [14,34,60]. These verification methods could play a significant role in the future certification of DNN-based systems.

Here, we deal with the case where we already know that a given DNN is malfunctioning; specifically, we assume we have a finite set of concrete inputs which are handled erroneously (discovered by testing, verification, or any other method). In this situation, we would like to *modify* the network, so that it produces correct predictions for these inputs. A naïve approach for accomplishing this is to add these faulty inputs to the training set used to create the DNN, and then retrain it, but this is often too computationally expensive [24]. Also, retraining may change the network significantly, potentially introducing new bugs on inputs that were previously correctly handled. Finally, retraining might be impossible when the original training set is inaccessible, e.g., due to its privacy or sensitivity [28].

Instead, we advocate an approach that requires no retraining, and which has recently gained some attention [11,43,59,63]: we present a new tool, called 3M-DNN (*M*inimal, *M*ulti-layer *M*odifications for *DNNs*), which can directly find a modification to the network and correct the erroneous behavior. In this context, a modification means changing the networks *weights*—the set of real values that determine the DNN's output, and which are initially selected during training. Further, because we assume the original network is mostly correct, we seek to find a modification which is also *minimal*. The motivation is that such a change would maintain as much as possible of the network's behavior on other inputs. In other words, our goal is to improve the DNN's overall *accuracy*—the percentage of correctly handled inputs, which is normally measured with respect to a *test set* of examples—by improving its handling of problematic inputs, and without harming its handling of other inputs.

A DNN is, by definition, a *layered* artifact; and to the best of our knowledge, all previous work on finding minimal modifications to a DNN's weights focused on changing the weights of a single layer [11,20,59]. Intuitively, and as we later demonstrate, this significant restriction could prevent one from finding potentially smaller (and thus preferable) changes to the network. In 3M-DNN, we seek to lift this restriction by proposing and implementing a novel method for the *multi-layer* modification of a DNN, with the goal of finding smaller modifications than could be otherwise possible. The key idea of our approach is to split the network into multiple sub-networks along certain layers, which we refer to as *separation layers*; and then attempt to find a minimal change for each of these sub-networks separately, in a way that brings about the desired overall change to the network.

More concretely, 3M-DNN is comprised of two logical levels. In the top, *search level*, the tool conducts a heuristic search through possible changes to the values computed by the separation layers. Each possible change to these values that we consider, translates into a possible fix to the DNN; it naturally gives rise to a sequence of problems on the bottom, *single-layer modification level*, each involving a single sub-network. Solving these single-layer modification problems can be performed using existing techniques; and the changes discovered to the sub-networks modify the values of the separation layers as selected by the top level. Thus, the process as a whole allows 3M-DNN to reduce the problem of multi-layer changes into a sequence of single-layer change problems, which can be dispatched using existing DNN modification tools as backends.

In its search for a minimal change, 3M-DNN alternates between the two levels: each time the top-level examines a potential change to the separation layers, and invokes the lower level in order to compute the overall cost of using that change (by combining the costs of changing each individual sub-network). The top-level always maintains the minimal change it has encountered so far, and uses search heuristics in order to find new, better options. The search space is infinite, and so our tool is *anytime*—it is designed to be run with a timeout, and whenever it is stopped, it returns the best (smallest) change discovered so far.

The search heuristic used by the top-level can have a crucial impact on performance. The approach implemented in 3M-DNN is general, in the sense that any search heuristic can be plugged in; and here, we consider and implement three such heuristics. The first is a random search, in which the top level randomly explores possible changes; this heuristic serves as a baseline. The second is a greedy search heuristic, in which the search always progresses in the direction that produces the most immediate gain. The third heuristic is a Monte Carlo Tree Search (MCTS) approach [7], which attempts to balance between exploration of the search space and the exploitation of regions already known to produce good solutions.

The 3M-DNN tool is available online.[1] It is designed in a modular fashion, so that additional search heuristics can be plugged in; it currently uses the Marabou DNN verification tool [34,55,61,62] as a backend, although other tools could be used as well. We used 3M-DNN to compare the different aforementioned heuristic strategies, and to compare our method to a single-layer modification method, with respect to the accuracy and minimal change size found. In our experiments, 3M-DNN achieved favorable results when compared to single-layer modification techniques. The greedy and MCTS heuristics both performed better than the random one; and while the greedy approach generally outperformed MCTS, there were cases where the latter proved superior. Finally, we also used 3M-DNN to find three-layers modification to a network, as a proof-of-concept that demonstrates its ability to modify any number of layers simultaneously.

The rest of this paper is organized as follows. In Sect. 2 we provide the necessary background on DNNs and repairing DNNs with minimal modifications. In Sect. 3 we describe 3M-DNN's algorithm for multi-layer modification in greater detail, and explain its different strategies for the heuristic search. Then, in Sect. 4 we provide additional technical details on our implementation of 3M-DNN. We describe our experiments and results in Sect. 5. In Sect. 6 we review relevant related work, and finally in Sect. 7 we conclude and describe our plans for future work.

---

[1] https://zenodo.org/record/5735194#.Ysvf_nZByUk.

2   Background

Deep Neural Networks. A deep neural network (a model) N is comprised of n layers, $L_1, \ldots, L_n$. Each layer L_i is comprised of s_i nodes, also called *neurons*. The first layer, L_1, is the *input layer*, and is used to provide the network with an input vector $v_1 \in \mathbb{R}^{s_1}$. The network is then evaluated by iteratively computing the assignment v_i of layer L_i for $i = 2, \ldots, n$, each time using the assignment v_{i-1} as part of the computation. Finally, the DNN computes the assignment v_n of layer L_n, which is the *output layer*. v_n serves as the output of the entire neural network. Layers $L_2, \ldots, L_{n-1}$ are referred to as *hidden layers*.

Each assignment v_i for $2 \leq i \leq n$ is computed by multiplying v_{i-1} by a real-valued *weight matrix* θ_i, and applying a non-linear *activation function* (except for the final output layer, where no activation function is applied). We use θ to denote the set of all weights $\theta = [\theta_2, \ldots, \theta_n]$, and use N_θ to refer to the function $N_\theta : \mathbb{R}^{s_1} \to \mathbb{R}^{s_n}$ computed by N. The weight matrices θ_i are key, and they are selected during the network's *training phase*, which is beyond our scope here (see, e.g., [21] for details). Modern DNNs use various activation functions [47]; for simplicity, we restrict our attention here to the popular *rectified linear unit* (ReLU) function, defined as

$$\mathrm{ReLU}(x) = \max\,(0, x),$$

although our approach could be used with other functions as well. When ReLUs are used, the values v_i of layer L_i are computed as $v_i = \mathrm{ReLU}(\theta_i \cdot v_{i-1})$, where the ReLUs are applied element-wise. We use the term *network architecture* to refer to the number of layers in N, the size of each layer s_i, and the activation functions in use. Note that the network's weights are not considered part of the network's architecture.

For a given point $x \in \mathbb{R}^{s_1}$, we refer to the assignment of the output layer $N_\theta(x)$ as the network's *prediction* on x. A common class of DNNs are designed for the purpose of *classification*, where the maximal entry of the prediction $N_\theta(x)$ indicates the *label* to which x is classified. In other words, the *classification* of $x \in \mathbb{R}^{s_1}$ as determined by N_θ is defined as $\arg\max N_\theta(x)$. Classification DNNs are useful, for example, for image recognition [51], and are highly popular. When dealing with classification networks, we say that N_θ produces an erroneous output for x if it classifies it differently than some given, ground-truth label l:

$$\arg\max N_\theta(x) \neq l$$

A small, running example is depicted in Fig. 1. This toy DNN is comprised of five layers—an input layer with a single node, three hidden layers with two nodes each, and an output layer with two nodes. The weight of each edge appears in the figure (a missing edge indicates a weight of 0). All activation functions in this example are ReLUs. When the network is evaluated on input $v_1 = [1]$, the assignment of the first hidden layer is $v_2 = [1, 1]$; the second hidden layer evaluates to $v_3 = [0.01, 100]$; the third hidden layer evaluates to $v_4 = [10, 1]$; and finally, the output layer evaluates to $v_5 = [11, -11]$. If we treat this DNN as a classification model, the classification of $x = 1$ is 1, as $11 = v_5^1 > v_5^2 = -11$.

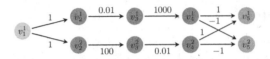

Fig. 1. A toy DNN.

Repairing DNNs with Minimal Modification. For a given DNN N_θ : $\mathbb{R}^{s_1} \to \mathbb{R}^{s_n}$ with n layers, and a finite set of points $S \subset \mathbb{R}^{s_1}$ for which we know N_θ produces a wrong prediction, our goal is to change the network's weights θ, so that its classification of S becomes correct.

We begin by formally defining the *minimal modification problem* for classification networks (later, we extend this definition to other networks as well). Let N_θ be a classification DNN, let S be a set of inputs, and let F be an oracle function $F : \mathbb{R}^{s_1} \to \{1, \ldots, s_n\}$ which indicates the correct classification for each point $x \in S$. Our goal is to produce a modification to θ, which we denote δ, and obtain a new set of weights $\theta' = \theta + \delta$, such that:

$$\forall x \in S. \quad \arg\max N_{\theta'}(x) = F(x) \tag{1}$$

Observe that the architecture of N is unchanged. Our goal is to find a δ that is *minimal*, with the goal of preserving N's behavior on points outside S. The magnitude of δ can be measured using any metric, such as the L_1 or L_∞ norms.

Using these definitions, the *minimal modification problem* for classification DNNs is defined as follows:

Definition 1. *The Minimal Modification Problem for Classification Models.* *Let $N_\theta : \mathbb{R}^{s_1} \to \mathbb{R}^{s_n}$ be a classification model with n layers, and let $S \subset \mathbb{R}^{s_1}$ be a set of points. Let $F : S \to \{1, \ldots, s_n\}$ be an oracle function, which indicates the correct classification for each $x \in S$. Let $\|.\|$ be some norm function. The Minimal Modification Problem is:*

$$minimize \ \|\delta\|$$
$$subject \ to \ \arg\max N_{\theta'}(x) = F(x) \quad \forall x \in S$$
$$\theta' = \theta + \delta$$

We continue with our running example from Fig. 1. Recall that for input $x = 1$, we get $v_5^1 = 11$ and $v_5^2 = -11$. Now assume that $S = \{1\}$, and that the desired classification for $x = 1$ is actually $F(1) = 2$. Thus, we need the network to satisfy that $v_5^1 < v_5^2$ when evaluated on $x = 1$. We make an even stronger requirement, that $v_5^1 + \mu \le v_5^2$, for some small $\mu > 0$; this guarantees a small gap in the scores assigned to v_5^1 and v_5^2, and avoids draws. For this example, we set $\mu = 0.1$. Using the L_1 norm, the minimal single-layer modification that achieves the desired changes has size 2.21, as depicted in Fig. 2. With this change to the network, we get that $v_5^1 = -11.1 < -11 = v_5^2$. However, if we allow changing two layers, we can actually achieve a *smaller* minimal modification of size 2.11,

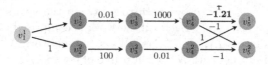

Fig. 2. Minimal single-layer modification for the toy example of Fig. 1. The only changed layer is the output layer, where the weight of the edge $v_4^1 \to v_5^1$ was changed from 1 to -1.21. The size of the change (using the L_1 norm) is 2.21.

which is preferable because it has a smaller impact on the DNN's behavior. We will later return to this example in Sect. 3.

Definition 1 is typically sufficient for classification DNNs, but it can be generalized to support arbitrary constraints on the DNN's outputs. Let N_θ be a general DNN (not necessarily a classification DNN). For each point $x \in S$, we consider a matrix $A_x \in \mathbb{R}^{k_x \times s_n}$ and a vector $b_x \in \mathbb{R}^{k_x}$, where k_x is the number of linear constraints on the output of the network on x. The aim is to produce a modification to θ, which we denote again with δ, and get new weights $\theta' = \theta + \delta$, which satisfies:

$$A_x N_{\theta'}(x) \le b_x \qquad (2)$$

Under this formulation we can express constraints such as "the first output of $N_{\theta'}$ on x should satisfy $3 \le N_{\theta'}(x) \le 5$", which could not be expressed using the previous formulation. This formulation subsumes the classification case. Again notice that we keep the architecture of N the same, and we only make modifications to θ. More formally, the minimal modification problem for the general case is defined as follows:

Definition 2. The Minimal Modification Problem. *Let $N_\theta : \mathbb{R}^{s_1} \to \mathbb{R}^{s_n}$ be a DNN model with n layers, and let $S \subset \mathbb{R}^{s_1}$ be a set of points. For each point $x \in S$, let $A_x \in \mathbb{R}^{k_x \times s_n}$, $b_x \in \mathbb{R}^{k_x}$ be the output constraints of N_θ on x. Let $\|.\|$ be some norm function. The Minimal Modification Problem is:*

$$minimize \ \|\delta\|$$
$$subject \ to \ A_x N_{\theta'}(x) \le b_x \quad \forall x \in S$$
$$\theta' = \theta + \delta$$

To the best of our knowledge, all previous approaches for solving the problems stated in Definitions 1 and 2 focused on finding a minimal modification for only a single layer of N. In contrast, in 3M-DNN we seek to solve the problem while allowing multiple layers of N to be modified, as we discuss next.

3   3M-DNN: Finding Multi-Layer DNN Changes

The key idea incorporated into 3M-DNN is to reduce the multi-layer modification problem into a sequence of single-layer modification problems. Specifically, given a DNN N with n layers $L_1, \ldots, L_n$ and a list of k separation layer indices $1 < i_1 < \ldots < i_k < n$, we wish to partition the layers of N into $k + 1$ sub-networks $N^0, N^1, \ldots, N^k$. Each sub-network is comprised of a subset of the original network's layers $L_1, \ldots, L_n$, as follows: sub-network N^0 is comprised of layers $L_1, \ldots, L_{i_1}$; sub-network N^k is comprised of layers $L_{i_k}, \ldots L_n$; and for each $1 \leq j \leq k - 1$, sub-network N^j is comprised of layers $L_{i_j}, \ldots, L_{i_{j+1}}$. Note that each pair of consecutive sub-networks N^j and N^{j+1} both contain layer $L_{i_{j+1}}$, which functions once as N^j's output layer, and once as N^{j+1}'s input layer. We refer to the shared layers $L_{i_1}, \ldots, L_{i_k}$ as the *separation layers*.

We apply this partitioning to our running example, as depicted in Fig. 3. There, we split the DNN into two sub-networks N^0 and N^1, with the original L_3 layer serving as the only separation layer. Observe that the input layer of N^0 is the input layer of the original network, and that the output layer of N^1 is the output layer of the original network. Indeed, if we were to evaluate N^0 on some input x, and then feed its output as the input to N^1, then N^1's output would match the output of the original network when evaluated on x.

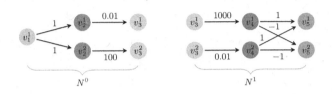

Fig. 3. Splitting a network along a separation layer.

Next, we wish to modify N^0 and N^1, and then combine these modifications into a modification of the original network. Let $S = \{1\}$, i.e. $x = 1$ is our only misclassified input, and let us require that x be classified as class 2. In other words, we wish $N(1)$ to produce output values for which $v_5^1 + \mu \leq v_5^2$ for some small $\mu > 0$. 3M-DNN begins by deciding on a change of values for the neurons of the separation layer, v_3^1 and v_3^2. In the original evaluation of the network on $x = 1$, we got $v_3^1 = 0.01$ and $v_3^2 = 100$. Let us require that v_3^1's value be changed to 0, and that v_3^2's value remains unchanged. This requirement translates into two single-layer modification queries: for N^0, 3M-DNN will require that on input $x = 1$, the outputs be $[0, 100]$; and for N^1, 3M-DNN will require that on input $[0, 100]$, the network's outputs satisfy $v_5^1 + \mu \leq v_5^2$. Both these single-layer modification queries can be solved using a black-box modification procedure; for example, here, if we assume again that $\mu = 0.1$, a possible modification is to change the weight of edge $v_2^1 \to v_3^1$ to 0 in N^0, and to change the weight of edge $v_4^2 \to v_5^2$ to 1.1 in N^1. Applying both of these changes to the original network produces a modification of size 2.11 (using the L_1-norm), which results in the desired behavior for $x = 1$; indeed, after applying this change, we get that $1 = v_5^1 < v_5^2 = 1.1$. The modified network is depicted in Fig. 4. Observe that this change is minimal for our particular selection of a separation layer index and the ensuing selection of changes to the separation layer's assignment; but it is not necessarily globally minimal, as a different choice of separation index or assignment could result in smaller changes.

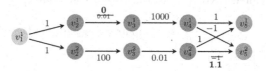

Fig. 4. The two-layer modification found using 3M-DNN.

The example described above is generalized into 3M-DNN's full algorithm, which appears as Algorithm 1. For simplicity of presentation, Algorithm 1 handles the classification model case from Definition 1; 3M-DNN actually supports the more general case from Definition 2, and the implemented algorithm is very similar to the one given here. Algorithm 1 takes as input the DNN N in question, the set of misclassified points S and the oracle function F that describes these points desired classification; the separation indices $I = \{i_1, \ldots, i_k\}$ indicating how the network is to be broken down into sub-networks, in which only a single layer will be changed; and a timeout value T. The algorithm then begins its heuristic search for a minimal change to the network that brings about the desired changes.

First, in Lines 1–3, the algorithm evaluates the assignments of the separation layers, for each input point in S. Then, in Line 4, the algorithm constructs the sub-networks $N^0, \ldots, N^k$, according to the separation indices. Recall that our algorithm is anytime, i.e., always maintains the best modification discovered so far; this modification, and its cost (i.e., its distance from the original network according to the distance metric in use) is stored in the variables initialized in Line 5. The algorithm then begins running in a loop until exhausting its timeout value.

In every iteration of its main loop, the algorithm begins (Lines 8–11) by selecting a modified assignment for each separation layer L_{i_l} for $1 \leq l \leq k$. This modification is selected by the place-holder function PROPOSECHANGE(); this function is where the heuristic search used in the search level of 3M-DNN comes into play. We discuss these heuristics in detail in Sect. 3.1. Then, in Lines 13–17, the algorithm computes for each of the sub-networks $N^0, \ldots, N^k$ the minimal, single-layer changes required to bring about the global changes selected by the search level. These changes are computed by repeated invocations of the SINGLELAYERMODIFICATION() function, which is again a place-holder function that represents the single-layer modification level of 3M-DNN; we describe it in more depth in Sect. 3.2. This function takes as input a DNN, and a list of pairs of input points and their desired outputs; and returns the modified DNN, and

---

Algorithm 1 The 3M-DNN Algorithm (For Classification Networks)

Input: DNN N, set of input points $S = \{x_1, \ldots, x_n\}$, oracle function F, separation indices $I = \{i_1, \ldots, i_k\}$, timeout T

Output: A repaired DNN N' with the same architecture as N

1: **for** $j = 1 \ldots n$ **do**
2:     $v_{i_1}^j, \ldots, v_{i_k}^j \leftarrow N(x_j)$ ▷ Compute the separation layers' assignments
3: **end for**
4: $N^0, \ldots, N^k \leftarrow \text{SPLIT}(N, I)$
5: bestChange $\leftarrow \bot$, bestCost $\leftarrow \infty$
6: **while** timeout T not exceeded **do**
7:     **for** $l = 1 \ldots k$ **do**
8:         $c_l \leftarrow \text{PROPOSECHANGE}()$
9:         **for** $j = 1 \ldots n$ **do**
10:             $v_{i_l}^{'j} \leftarrow v_{i_l}^j + c_l$ ▷ Select new assignments for the separation layers
11:         **end for**
12:     **end for**
13:     $N'^0, cost_0 \leftarrow \text{SINGLELAYERMODIFICATION}(N^0, \langle x_1, v_{i_1}^{'1} \rangle, \ldots, \langle x_n, v_{i_1}^{'n} \rangle)$
14:     **for** $l = 1 \ldots k - 1$ **do**
15:         $N'^l, cost_l \leftarrow \text{SINGLELAYERMODIFICATION}(N'^l, \langle v_{i_l}^{'1}, v_{i_{l+1}}^{'1} \rangle, \ldots, \langle v_{i_l}^{'n}, v_{i_{l+1}}^{'n} \rangle)$
16:     **end for**
17:     $N'^k, cost_k \leftarrow \text{SINGLELAYERMODIFICATION}(N^k, \langle v_{i_k}^{'1}, F(x_1) \rangle, \ldots, \langle v_{i_k}^{'n}, F(x_n) \rangle)$
18:     cost $\leftarrow \text{TOTALCOST}(cost_0, \ldots, cost_k)$
19:     **if** cost < bestCost **then**
20:         bestCost $\leftarrow$ cost
21:         bestChange $\leftarrow \langle N'^0, \ldots, N'^k \rangle$
22:     **end if**
23: **end while**
24: **return** $\langle$ bestCost, $\text{COMBINE}(\text{bestChange}) \rangle$

---

the modification's cost.[2] In Line 13, we use SINGLELAYERMODIFICATION() to modify N^0: we required that the original input points $x_1, \ldots, x_n$ produce outputs that match the selected modified assignments $v_{i_1}^{'1}, \ldots, v_{i_1}^{'n}$ of L_1. In Line 15, SINGLELAYERMODIFICATION() is used to modify each of the $N^1, \ldots, N^{k-1}$ subnetworks, so that each sub-network produces as output the input selected for its successor. Finally, in Line 17, the last sub-network N^k is modified, so that it produces outputs that match the oracle's predictions on the original input points.

The single-layer modification procedures invoked for $N^0, \ldots, N^k$ each return the modified sub-networks $N'^0, \ldots, N'^k$, and the respective costs of the modifications $cost_0, \ldots, cost_k$. The total modification cost for the complete network is then computed by the TOTALCOST() function in Line 18, whose implementation

---

[2] It may be possible that an invocation of SINGLELAYERMODIFICATION() fails because no change is possible that obtains the desired results. Whenever this happens, 3M-DNN continues to the next iteration, exploring a different change to the separation layers' values. This situation is theoretically possible, but did not occur in our experiments.

depends on the norm used for measuring distance; for example, in the case of L_1 norm, it returns the sum of its inputs; for L_∞, it returns the maximal input; etc. The modified sub-networks with the lowest total cost found so far, along with the cost itself, are saved in Lines 19–22.

The algorithm halts when the provided timeout is exhausted, and it then returns the complete modified network with the best modifications found so far, and the cost of that modification. The re-assembling of the complete modified network is performed by the function COMBINE(), whose implementation is omitted for brevity.

Soundness and Completeness. Assuming that the SINGLELAYERMODIFICATION() is sound—for example, if it is implemented using a sound DNN verifier [20]—any modification returned by our tool will indeed correct the global DNN behavior on the input set S. In that sense, 3M-DNN is sound. It is, however, generally incomplete; there are infinitely many modifications that can be attempted for the separation layers, and it is infeasible to try them all. This is our motivation for introducing the timeout mechanism and making the algorithm anytime; and indeed, the algorithm is not guaranteed to return the smallest change possible. It does, however, attempt to minimize the change based on search heuristics that we discuss next.

3.1 The Search Level

Algorithm 1 considers an infinite space of possible changes to the values of the separation layers, each time selecting a single possible change and computing its cost (Line 8 of the Algorithm). For a single separation layer with n neurons, the search space is $\mathbb{R}^n$ in its entirety, and the problem is compounded when multiple separation layers are involved. To exacerbate matters even further, the computed cost function for possible changes need not be convex; see Fig. 5 for an illustration.

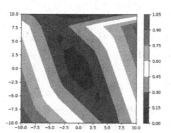

Fig. 5. The cost function for a small DNN, with a single separation layer with 2 neurons. The X and Y axes represents the change for each neuron, and the color represents the size of the minimal modification achieved. The function is not convex.

To circumvent this difficulty, we first define the following grid, parameterized by a step size ϵ:

$$\mathcal{G}_\epsilon = \{v = \langle \alpha_1 \cdot \epsilon, \alpha_2 \cdot \epsilon, \ldots, \alpha_n \cdot \epsilon \rangle \mid \alpha_i \in \mathbb{Z}\}$$

Each point in the grid represents a single, possible change for a separation layer, and the discretization allows us to better handle the search space. Naturally, this comes at the cost of possibly overlooking better changes that do not coincide with the grid, but this can be mitigated by making the grid dense (picking a smaller ϵ). The grid's origin, i.e., point $0^n \in \mathbb{R}^n$, corresponds to no change at all to the separation layer; and points that are very far away from the origin are likely to represent significant changes to the DNN.

Despite the discretization, the grid is still infinite and multi-dimensional, and so 3M-DNN implements three search heuristics: *random search, greedy search* and *Monte-Carlo Tree Search (MCTS)*. Each of these heuristics can be regarded as a possible implementation of the PROPOSECHANGE() method from Algorithm 1. We next elaborate on each of them.

Random Search. This heuristic performs a uniform random search over $\mathcal{G}_\epsilon$. Specifically, it samples a grid point uniformly at random, and that point constitutes the proposed change to the separation layer. We treat this simple heuristic as a baseline, to which the more sophisticated heuristics are compared.

Greedy Search. The motivation for this heuristic is that the optimal grid point is likely not far away from the origin (as far away points likely correspond to significant changes to the network). Thus, we start from the grid's origin as our current change, and at each iteration, consider the grid points that are immediate neighbors of our current points—that is, points obtained by adding or subtracting ϵ from one of the coordinates of the current point. We then compute the costs associated with each of these points, and pick the cheapest one as our new current point.

More formally, if $g_0 \in \mathcal{G}_\epsilon$ is our current search point, we observe all points $g \in \mathcal{G}_\epsilon$ such that $\|g_0 - g\|_{L_1} = \epsilon$, invoke the SINGLELAYERMODIFICATION() with appropriate parameters to compute the cost of each g, and update g_0 to be the g that obtained the lowest cost.

Monte Carlo Tree Search. The aforementioned greedy approach can be regarded as an attempt to optimize *exploitation*: whenever a good "direction" on the grid is discovered, we follow that direction. A natural concern is that such an approach might lead to local minima, and fail to detect cheaper changes that can only be reached via grid points with higher costs (recall that the cost function is not necessarily convex). To balance the greedy approach's exploitation with *exploration* for detecting possibly better changes, we employ a *Monte Carlo Tree Search (MCTS)* heuristic [7]. We give here a short overview of this approach; see [7] for a more in-depth review.

MCTS is a heuristic search algorithm over a discrete set of actions, with the goal of selecting the most promising move based on simulations. It has recently been shown quite successful in multiple application domains, most notably in board games such as Go [17]. The search is conducted on a *search tree*, where each node represents a state. The root node of the search tree represents the initial state, and a child of a node represents another state that can be reached by performing a single action. Initially, the entire search tree is yet *unexplored*; and the algorithm iteratively explores additional parts thereof, one node in every iteration. In our setting, each node of the search tree is a grid point; and the possible moves include moving to one of the adjacent grid points (similarly to the greedy approach).

In each iteration, MCTS performs *simulations* in order to decide which unexplored node to visit next. Specifically, these simulations allow MCTS to compute a cost for each of the candidate nodes, and then pick the candidate with the lowest cost as the next node to visit.

More concretely, each MCTS iteration consists of 4 steps:

1. *Selection*: one of the nodes at each level in the explored portion of tree is selected, according to some policy, until reaching a leaf node. A common policy, also used in 3M-DNN, is the *upper confidence bound* (*UCB*) policy. The policy's details are beyond our scope here; see [7] for additional details.
2. *Expansion*: one of the unexplored children of the leaf node from Step 1 is selected randomly.
3. *Simulation*: one or more simulations are carried out for the node selected in Step 2. Each simulation explores deeper into the search sub-tree rooted at the new node until reaching a predefined tree depth, by picking a child randomly in each level of the sub-tree. When the simulation arrives at the last node, it computes a cost value that takes into account all the steps that led from the node picked at Step 2 to the final node that was reached.
4. *Backpropagation*: the cost computed in each simulation is back-propagated through all the nodes in the path leading back up to the root. Each node aggregates the costs of simulations of paths containing it, and the aggregated cost is used for Step 1 in the next iteration of MCTS.

After reaching a predefined number of iterations, the unexplored node that has obtained the lowest cost so far is chosen as the next move.

In our implementation of the MCTS search heuristic, every invocation of PROPOSECHANGE() for a given separation layer L_j runs the MCTS algorithm, which in turn performs a predefined number of sub-iterations. The root of the search tree represents the current change to the assignment of L_j, and a move to a child node represents a single step along the grid. Consequently, for each tree node of the search tree in the MCTS algorithm, there are $2s_j + 1$ child nodes (including the option to not take a step at all). The simulation step of MCTS includes, in our case, dispatching single-layer modification queries.

3.2   The Single-Layer Modification Level

As part of its operation, our algorithm needs to dispatch numerous queries of single-layer modifications in DNN (the SINGLELAYERMODIFICATION() calls in Algorithm 1). In each of these queries, the sub-network in question has specific inputs, for which certain output constraints need to hold—either the outputs need to classify the inputs as a certain label (for the last sub-network), or they need to take on exact, predetermined values (for all other sub-networks). Solving such queries has been studied before, and as part of our solution, we propose to use existing techniques and tools as a backend. In our implementation (described in greater detail later), we used the approach proposed by Goldberger et al. [20].

4   Implementation

We implemented Algorithm 1 and the aforementioned search heuristics in the new 3M-DNN tool. 3M-DNN is implemented as a Python 3.7.3 module, and uses TensorFlow-Keras 2.3 as a backend for representing DNNs. We attempted to design 3M-DNN in a modular fashion, in order to easily allow the future addition of new search heuristics in the search level, as well as additional backend engines for dispatching single-layer modification queries.

The main class of 3M-DNN is the abstract *NetworkCorrectionMethod* class. It defines the interfaces and methods that a subclass must implement in order to fit the mold defined by Algorithm 1. Specifically, the class defines the following methods:

__init__(DNN N, $[x_1, \ldots, x_n]$, $[o_1, \ldots, o_n]$): a constructor for the inheriting class. It takes as input a TensorFlow-Keras DNN, a list of input points as NumPy arrays, and a list of output constraints for each point. Each output constraint is a list of 2 items: a NumPy array A and a NumPy vector b, and the output y of the corresponding point should satisfy $Ay \leq b$ (per Definition 2).

correct_network(): the main entry point for the inheriting class, responsible for running the correction procedure for the DNN and constraints provided through the constructor. Its implementation depends on the heuristic search method and the single-layer modification method chosen. Returns *True* if a modification to the network was found, or *False* otherwise.

get_corrected_network(): this method is invoked after *correct_network()*, and returns the corrected network as a tensorflow-keras model.

get_minimal_change(): a method called after *correct_network()*, which returns the list of the changes found during the modification process, for each changed layer.

get_changed_layers(): a method called after *correct_network()*, which returns a list of layer indices of the layers changed during the modification process.

Our implementation of 3M-DNN includes multiple instantiations of the *NetworkCorrectionMethod* class that implement the heuristics defined in Sect. 3. Specifically, class *NetworkCorrectionTwoLayersUniform* implements the random search heuristic; the core of the implementation appears in the *correct_network()* method. Similarly, class *NetworkCorrectionTwoLayersGreedy* implements the greedy search approach; and its core is again in method *correct_network()*. Finally, the MCTS approach is implemented in classes *NetworkCorrectionTwoLayersTreeSearch* and *MCTS*. Class *MCTS* controls the various configurable parameters of the search, such as the step size, the number of simulations per iteration, and the maximal depth of the search tree. All three grid search heuristics are currently linked to the Marabou DNN verification as the single-layer change backend; this connection is implemented in class *MarabouRunner*.

5   Evaluation

Setup. We used 3M-DNN to evaluate the usefulness of our modification approach. Specifically, we experimented with a DNN trained on the MNIST dataset for digit recognition [42]. The dataset contains 70,000 handwritten digit images with 28 × 28 pixels, split into a training set of 60,000 images, and a test set of 10,000 images. We trained a network N comprised of 8 layers: an input layer of size 784 neurons, six hidden layers, each of size 20 neurons, and an output layer with ten neurons. The hidden layers all used the ReLU activation function. We then used network N to conduct three kinds of experiment (all conducted with the L_∞-norm): (i) comparing search heuristics: an experiment where we used 3M-DNN to find two-layer modifications for N, using each of the three heuristic search strategies discussed in Sect. 3; (ii) comparing multi-layer and single-layer modifications: here, we used 3M-DNN to search for repairs for N that modified either a single layer or two layers, in order to evaluate the necessity of modifying multiple layers; and (iii) three-layer repairs: we attempted to repair N by modifying three layers, to demonstrate 3M-DNN ability to repair the network by changing any number of layers. Below we provide additional information on each of the experiments, and their results are summarized in Table 1.

Table 1. Results of experiments. The 1-Layer search strategy stands for a single-layer modification process. Greedy-3 stands for three-layer-modification using the greedy heuristic search.

Exp. #	Search Strategy	Number of input points	Average Change	Minimal Change	Maximal Change	Average Accuracy	Minimal Accuracy	Maximal Accuracy
1	Random	1	0.1520	0.0615	0.4922	0.6865	0.1916	0.9308
	Greedy		0.0133	0.001	0.0566	0.943	0.7971	0.9576
	MCTS		0.0139	0.001	0.0566	0.943	0.7971	0.9576
	Random	2	0.197	0.0791	0.4775	0.6302	0.2563	0.9161
	Greedy		0.0463	0.0058	0.1435	0.9245	0.7417	0.9598
	MCTS		0.0478	0.0058	0.1484	0.9261	0.7398	0.9594
2	Greedy	1	0.0305	0.0029	0.1699	0.9397	0.5856	0.9565
	1-Layer		0.0307	0.0029	0.1875	0.9394	0.585	0.9562
	Greedy	2	0.0459	0.0039	0.2041	0.9178	0.3124	0.9576
	1-Layer		0.0464	0.0039	0.208	0.9163	0.3124	0.9576
3	Greedy-3	1	0.25097	0.25097	0.25097	0.886	0.886	0.886

Experiment 1: Comparing Search Heuristics. We used 3M-DNN in each of the three search method configurations, to solve: (i) 100 benchmarks where N was modified to repair its output on 1 input point; and (ii) another 100 benchmarks with repair on 2 input points. In all experiments, we split N into two sub-networks along its fourth hidden layer, with $\epsilon = 0.5$ as the grid parameter; and the timeout value was set to 1000 s seconds. In experiments involving two input points, we expedited the process by restricting changes solely to the final layer of each sub-network. The results are summarized in Table 1, and illustrated in Fig. 6. Both the Greedy and MCTS strategies significantly outperform the uniform random search heuristic, achieving higher accuracy and smaller change size. The Greedy and MCTS heuristics are relatively equal in their performance, with each strategy outperforming the other in some cases.

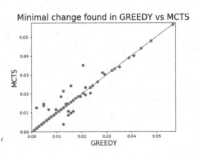

Fig. 6. Minimal modification size achieved by the Greedy and MCTS heuristic strategies in Experiment 1.

Experiment 2: Comparing Multi- And Single-Layer modifications.
Here, we configured 3M-DNN to use the greedy search heuristic, and used it
to solve: (i) 2000 minimal modification queries where where a single input point
had to be corrected; and (ii) another 2000 minimal modification queries with
repair on 2 input points. We ran each query once, looking for a one-layer minimal modification, and once searching for a two-layer modification. As before,
we set $\epsilon = 0.5$ and a timeout value of $1000\,s$ seconds (for both methods). To
expedite the experiments, we allowed the single-layer method to modify only
the final layer of the network [20], and the two-layers greedy method to modify
the last layer of each of the two sub-networks. Table 1 shows the superior performance of the two-layers greedy method over the single-layer method; although
the single-layer modification method was usually able to find its minimal modification within a minute, while the two-layers greedy method took longer. This
is not surprising, as the single-layer modification problem is significantly easier
computationally [20].

Experiment 3: Three-Layer Repairs. In the final experiment, serving as a
proof-of-concept, we used 3M-DNN to find a three-layer modification for N. We
ran this experiment once, with 3M-DNN configured to use the greedy search
heuristic on a single input point. We used a step size of $\epsilon = 0.5$. The timeout
value was set to $3600\,s$ seconds, and Table 1 depicts the results. The search space
when changing three layers is significantly more complex than in the previous
experiments, and so it is not surprising that 3M-DNN was only able to discover
changes that were larger than before. As we continue to improve our search
heuristics, and as the underlying verification engines continue to improve, the
scalability of 3M-DNN will also improve.

6   Related Work

The need to modify existing DNNs in order to correct them naturally arises
as part of the DNN life cycle, and has been a topic of interest in the wider
machine learning community. Most existing approaches are heuristic in nature:
for example, one approach is to iteratively apply Max-SMT solvers in search for
changes to the DNN [53]; another is to use reachability analysis to enrich the
training data [63]; and yet another approach is to heuristically identify "problematic" neurons and modify them [11]. A common property of most of these
approaches is that, in contrast to verification-based approaches, they provide no
formal guarantees about the minimality of the fixes that they produce.

Another approach for modifying the behavior of an existing DNN is to aug-
ment it with additional, non-DNN components that can override its output in
certain cases. This has been attempted using, e.g., decision trees [35,36] and
scenario objects [29,33]. A different technique is to transform the DNN into
another object, which is simpler to repair: for example, a pair of DNNs, in which
one determines the weights and another the activation functions [54]; or a DNN
with a self-repairing output layer [43]. Our technique is separated from these

approaches by the fact that the repaired artifact that it produces is a standard DNN, and is thus directly compatible with existing tools and infrastructure.

The approach that we take here, namely the application of DNN verification technology in order to find minimal modifications, has already received some attention. The approach that most closely resembles our own is the one proposed by Goldberger et al. [20]; and a related approach has also been proposed by Usman et al. [59]. However, these approaches are limited to modifying a single layer of the DNN in question, whereas the novelty of our approach is in enabling the simultaneous modification of multiple layers.

The technique proposed here uses a DNN verification engine as a black-box. DNN verification is an active research field, with many available tools and techniques. These include SMT-based approaches [25,30,32,34], LP- and MILP-solver based approaches [8,14,26,58], symbolic interval propagation [60], abstraction and abstract-interpretation based techniques [5,15,18,48,64], techniques for tackling recurrent networks [27,65] and binarized networks [4,46], and many others (e.g., [13,45,49]); and these techniques have been applied to multiple ends, such as DNN ensemble optimization [2], verifying adversarial robustness properties [9,18,23,31,40,58], verifying hybrid systems with DNN controllers [12,56], verifying DNNs that serve as controllers for congestion control systems [3,16,37] or robots [1], and DNN simplification [19,41]. As DNN verification engines continue to improve, so will the speed and scalability of our approach. Further, our line of work continues to demonstrate that DNN repair is an attractive application domain for verification.

7   Conclusion and Future Work

Due to the recent surge in DNN popularity, it is becoming increasingly important to provide tools and methodologies for facilitating tasks that naturally arise as part of DNN usage—such as modifying existing DNNs. Verification-based modification techniques offer significant advantages, and in this work, we have taken a step towards improving their applicability. Specifically, we were able to move beyond the single-layer change barrier that existed in prior work, and propose an approach that can simultaneously modify multiple layers of the DNN. Consequently, our approach can find modifications that are superior to those that would have been discovered by existing techniques.

Moving forward, we plan to extend our approach along several axes. First, we intend to explore additional strategies for conducting the grid search, as the strategy in use has a significant effect on overall performance. Specifically, we intend to train a *DNN controller* to manage the search strategy. Second, we observe that the grid search naturally lends itself to parallelization, and so we intend to explore parallelization techniques; and third, we intend to further demonstrate the usefulness of our technique by applying it to additional DNNs and case studies.

Acknowledgement. This work was partially supported by the Israel Science Foundation (grant number 683/18) and the HUJI Federmann Cyber Security Center.

References

1. Amir, G., et al.: Verifying Learning-Based Robotic Navigation Systems. Technical report (2022). http://arxiv.org/abs/2205.13536
2. Amir, G., Katz, G., Schapira, M.: Verification-aided deep ensemble selection. In: Proceedings of 22nd International Conference on Formal Methods in Computer-Aided Design (FMCAD) (2022)
3. Amir, G., Schapira, M., Katz, G.: Towards scalable verification of deep reinforcement learning. In: Proceedings of 21st International Conference on Formal Methods in Computer-Aided Design (FMCAD), pp. 193–203 (2021)
4. Amir, G., Wu, H., Barrett, C., Katz, G.: An SMT-based approach for verifying binarized neural networks. In: Proceedings of 27th International Conference on Tools and Algorithms for the Construction and Analysis of Systems (TACAS), pp. 203–222 (2021)
5. Ashok, P., Hashemi, V., Kretinsky, J., Mohr, S.: DeepAbstract: neural network abstraction for accelerating verification. In: Proceedings of 18th International Symposium on Automated Technology for Verification and Analysis (ATVA), pp. 92–107 (2020)
6. Bojarski, M., et al.: End to End Learning for Self-Driving Cars. Technical report (2016). http://arxiv.org/abs/1604.07316
7. Browne, C., et al.: A survey of monte Carlo tree search methods. IEEE Trans. Comput. Intell. AI Games 4(1), 1–43 (2012)
8. Bunel, R., Turkaslan, I., Torr, P., Kohli, P., Mudigonda, P.: A Unified View of Piecewise Linear Neural Network Verification. In: Proceedings of 32nd Conference on Neural Information Processing Systems (NeurIPS), pp. 4795–4804 (2018)
9. Carlini, N., Katz, G., Barrett, C., Dill, D.: Provably Minimally-Distorted Adversarial Examples. Technical report (2017). http://arxiv.org/abs/1709.10207
10. Ciregan, D., Meier, U., Schmidhuber, J.: Multi-column deep neural networks for image classification. In: Proceedings of IEEE Conference on Computer Vision and Pattern Recognition (CVPR), pp. 3642–3649 (2012)
11. Dong, G., Sun, J., Wang, J., Wang, X., Dai, T.: Towards Repairing Neural Networks Correctly. Technical report (2020). http://arxiv.org/abs/2012.01872
12. Dutta, S., Chen, X., Sankaranarayanan, S.: Reachability analysis for neural feedback systems using regressive polynomial rule inference. In: Proceedings of 22nd ACM International Conference on Hybrid Systems: Computation and Control (HSCC) (2019)
13. Dutta, S., Jha, S., Sanakaranarayanan, S., Tiwari, A.: Output range analysis for deep neural networks. In: Proceedings of 10th NASA Formal Methods Symposium (NFM), pp. 121–138 (2018)
14. Ehlers, R.: Formal verification of piece-wise linear feed-forward neural networks. In: Proceedings of 15th International Symposium on Automated Technology for Verification and Analysis (ATVA), pp. 269–286 (2017)
15. Elboher, Y., Gottschlich, J., Katz, G.: An abstraction-based framework for neural network verification. In: Proceedings of 32nd International Conference on Computer Aided Verification (CAV), pp. 43–65 (2020)
16. Eliyahu, T., Kazak, Y., Katz, G., Schapira, M.: Verifying learning-augmented systems. In Proceedings of Conference on the ACM Special Interest Group on Data Communication on the Applications, Technologies, Architectures, and Protocols for Computer Communication (SIGCOMM), pp. 305–318 (2021)

17. Fu, M.: AlphaGo and Monte Carlo tree search: the simulation optimization perspective. In: Proceedings of Winter Simulation Conference (WSC), pp. 659–670 (2016)
18. Gehr, T., Mirman, M., Drachsler-Cohen, D., Tsankov, E., Chaudhuri, S., Vechev, M.: AI2: safety and robustness certification of neural networks with abstract interpretation. In: Proceedings of 39th IEEE Symposium on Security and Privacy (S&P) (2018)
19. Gokulanathan, S., Feldsher, A., Malca, A., Barrett, C., Katz, G.: Simplifying neural networks using formal verification. In: Proceedings of 12th NASA Formal Methods Symposium (NFM), pp. 85–93 (2020)
20. Goldberger, B., Adi, Y., Keshet, J., Katz, G.: Minimal modifications of deep neural networks using verification. In: Proceedings of 23rd International Conference on Logic for Programming, Artificial Intelligence and Reasoning (LPAR), pp. 260–278 (2020)
21. Goodfellow, I., Bengio, Y., Courville, A., Bengio, Y.: Deep Learning. MIT Press Cambridge, Cambridge (2016)
22. Goodfellow, I., Shlens, J., Szegedy, C.: Explaining and Harnessing Adversarial Examples. Technical report (2014). http://arxiv.org/abs/1412.6572
23. Gopinath, D., Katz, G., Păsăreanu, C., Barrett, C.: DeepSafe: a data-driven approach for checking adversarial robustness in neural networks. In: Proceedings of 16th International Symposium on Automated Technology for Verification and Analysis (ATVA), pp. 3–19 (2018)
24. Hao, K.: Training a Single AI Model can Emit as much Carbon as Five Cars In Their Lifetimes. MIT Technology Review (2019)
25. Huang, X., Kwiatkowska, M., Wang, S., Wu, M.: Safety verification of deep neural networks. In: Proceedings of 29th International Conference on Computer Aided Verification (CAV), pp. 3–29 (2017)
26. Isac, O., Barrett, C., Zhang, M., Katz, G.: Neural network verification with proof production. In: Proceedings of 22nd International Conference on Formal Methods in Computer-Aided Design (FMCAD) (2022)
27. Jacoby, Y., Barrett, C., Katz, G.: Verifying recurrent neural networks using invariant inference. In: Proceedings of 18th International Symposium on Automated Technology for Verification and Analysis (ATVA), pp. 57–74 (2020)
28. Julian, K., Lopez, J., Brush, J., Owen, M., Kochenderfer, M.: Policy compression for aircraft collision avoidance systems. In: Proceedings of 35th Digital Avionics Systems Conference (DASC), pp. 1–10 (2016)
29. Katz, G.: Guarded deep learning using scenario-based modeling. In: Proceedings of 8th International Conference on Model-Driven Engineering and Software Development (MODELSWARD), pp. 126–136 (2020)
30. Katz, G., Barrett, C., Dill, D., Julian, K., Kochenderfer, M.: Reluplex: an efficient SMT solver for verifying deep neural networks. In: Proceedings of 29th International Conference on Computer Aided Verification (CAV), pp. 97–117 (2017)
31. Katz, G., Barrett, C., Dill, D., Julian, K., Kochenderfer, M.: Towards proving the adversarial robustness of deep neural networks. In: Proceedings of 1st Workshop on Formal Verification of Autonomous Vehicles (FVAV), pp. 19–26 (2017)
32. Katz, G., Barrett, C., Dill, D., Julian, K., Kochenderfer, M.: Reluplex: a calculus for reasoning about deep neural networks. In: Formal Methods in System Design (FMSD) (2021)
33. Katz, G., Elyasaf, A.: Towards combining deep learning, verification, and scenario-based programming. In: Proceedings of 1st Workshop on Verification of Autonomous and Robotic Systems (VARS), pp. 1–3 (2021)

34. Katz, G., et al.: The Marabou framework for verification and analysis of deep neural networks. In: Proceedings of 31st International Conference on Computer Aided Verification (CAV), pp. 443–452 (2019)
35. Kauschke, D., Lehmann, S.: Towards Neural Network Patching: Evaluating Engagement-Layers and Patch-Architectures. Technical report (2018). http://arxiv.org/abs/1812.03468
36. Kauschke, S., Furnkranz, J.: Batchwise patching of classifiers. In: Proceedings of 32nd AAAI Conference on Artificial Alliance (2018)
37. Kazak, Y., Barrett, C., Katz, G., Schapira, M.: Verifying deep-RL-driven systems. In: Proceedings of 1st ACM SIGCOMM Workshop on Network Meets AI & ML (NetAI), pp. 83–89 (2019)
38. Kermany, D., et al.: Identifying medical diagnoses and treatable diseases by image-based deep learning. Cell 172(5), 1122–1131 (2018)
39. Krizhevsky, A., Sutskever, I., Hinton, G.: ImageNet classification with deep convolutional neural networks. In: Proceedings of 26th Conference on Neural Information Processing Systems (NIPS), pp. 1097–1105 (2012)
40. Kuper, L., Katz, G., Gottschlich, J., Julian, K., Barrett, C., Kochenderfer, M.: Toward Scalable Verification for Safety-Critical Deep Networks. Technical report (2018). http://arxiv.org/abs/1801.05950
41. Lahav, O., Katz, G.: Pruning and slicing neural networks using formal verification. In: Proceedings of 21st International Conference on Formal Methods in Computer-Aided Design (FMCAD), pp. 183–192 (2021)
42. LeCun, Y.: The MNIST Database of Handwritten Digits (1998). http://yann.lecun.com/exdb/mnist/
43. Leino, K., Fromherz, A., Mangal, R., Fredrikson, M., Parno, B., Păsăreanu, C.: Self-Repairing Neural Networks: Provable Safety for Deep Networks via Dynamic Repair. Technical report (2021). http://arxiv.org/abs/2107.11445
44. Liu, Y., et al.: Trojaning Attack on Neural Networks (2017)
45. Lomuscio, A., Maganti, L.: An Approach to Reachability Analysis for Feed-Forward ReLU Neural Networks. Technical report (2017). http://arxiv.org/abs/1706.07351
46. Narodytska, N., Kasiviswanathan, S., Ryzhyk, L., Sagiv, M., Walsh, T.: Verifying Properties of Binarized Deep Neural Networks. Technical report (2017). http://arxiv.org/abs/1709.06662
47. Nwankpa, C., Ijomah, W., Gachagan, A., Marshall, S.: Activation Functions: Comparison of Trends in Practice and Research for Deep Learning. Technical report (2018). http://arxiv.org/abs/1811.03378
48. Ostrovsky, M., Barrett, C., Katz, G.: An Abstraction-refinement approach to verifying convolutional neural networks. In: Proceedings of 20th International Symposium on Automated Technology for Verification and Analysis (ATVA) (2022)
49. Ruan, W., Huang, X., Kwiatkowska, M.: Reachability analysis of deep neural networks with provable guarantees. In: Proceedings of 27th International Joint Conference on Artificial Intelligence (IJCAI) (2018)
50. Seide, F., Li, G., Yu, D.: Conversational speech transcription using context-dependent deep neural networks. In: Proceedings of 12th Conference of the International Speech Communication Association (Interspeech), pp. 437–440 (2011)
51. Simonyan, K., Zisserman, A.: Very Deep Convolutional Networks for Large-Scale Image Recognition. Technical report (2014). http://arxiv.org/abs/1409.1556
52. Singh, S., Kumar, A., Darbari, H., Singh, L., Rastogi, A., Jain, S.: Machine translation using deep learning: an overview. In: Proceedings of International Conference on Computer, Communications and Electronics (Comptelix), pp. 162–167 (2017)

53. Sotoudeh, M., Thakur, A.: Correcting deep neural networks with small, generalizing patches. In: Workshop on Safety and Robustness in Decision Making (2019)

54. Sotoudeh, M., Thakur, A.: Provable repair of deep neural networks. In: Proceedings of 42nd ACM SIGPLAN International Conference on Programming Language Design and Implementation (PLDI), pp. 588–603 (2021)

55. Strong, C., et al.: Global optimization of objective functions represented by ReLU networks. J. Mach. Learn. **2021**, 1–28 (2021). https://doi.org/10.1007/s10994-021-06050-2 ¡error l="305" c="Invalid ¡error l="303" c="Invalid
command: paragraph not started." /¿ command: paragraph not started." /¿ ¡error l="305" c="Invalid
command: paragraph not started." /¿

56. Sun, X., Khedr, H., Shoukry, Y.: Formal verification of neural network controlled autonomous systems. In: Proceedings of 22nd ACM International Conference on Hybrid Systems: Computation and Control (HSCC) (2019)

57. Sun, Y., Huang, X., Kroening, D., Sharp, J., Hill, M., Ashmore, R.: Testing Deep Neural Networks. Technical report (2018). http://arxiv.org/abs/1803.04792

58. Tjeng, V., Xiao, K., Tedrake. Evaluating robustness of neural networks with mixed integer programming. In: Proceedings of 7th International Conference on Learning Representations (ICLR) (2019)

59. Usman, M., Gopinath, D., Sun, Y., Noller, Y., Păsăreanu, C.: NNrepair: Constraint-based Repair of Neural Network Classifiers. Technical report (2021). http://arxiv.org/abs/2103.12535

60. Wang, S., Pei, K., Whitehouse, J., Yang, J., Jana, S.: Formal security analysis of neural networks using symbolic intervals. In: Proceedings of 27th USENIX Security Symposium, pp. 1599–1614 (2018)

61. Wu, H., et al.: Parallelization techniques for verifying neural networks. In: Proceedings of 20th International Conference on Formal Methods in Computer-Aided Design (FMCAD), pp. 128–137 (2020)

62. Wu, H., Zeljić, A., Katz, G., Barrett, C.: Efficient neural network analysis with sum-of-infeasibilities. In: Proceedings of 28th International Conference on Tools and Algorithms for the Construction and Analysis of Systems (TACAS), pp. 143–163 (2022)

63. Yang, X., Yamaguchi, T., Tran, H.-D., Hoxha, B., Johnson, T., Prokhorov, D.: Neural Network Repair with Reachability Analysis. Technical report (2021). http://arxiv.org/abs/2108.04214

64. Zelazny, T., Wu, C., Barrett, H., Katz, G.: On reducing over-approximation errors for neural network verification. In: Proceedings of 22nd International Conference on Formal Methods in Computer-Aided Design (FMCAD) (2022)

65. Zhang, H., Shinn, M., Gupta, A., Gurfinkel, A., Le, N., Narodytska, N.: Verification of recurrent neural networks for cognitive tasks via reachability analysis. In: Proceedings of 24th Conference of European Conference on Artificial Intelligence (ECAI) (2020)

Differentiable Logics for Neural Network Training and Verification

Natalia Ślusarz[(✉)], Ekaterina Komendantskaya, Matthew L. Daggitt,
and Robert Stewart

Heriot-Watt University,Edinburgh, UK
{nds1,ek19,md2006,rs46}@hw.ac.uk

Abstract. Neural network (NN) verification is a problem that has drawn attention of many researchers. The specific nature of neural networks does away with the conventional assumption that a static program is given for verification as in the case of NNs multiple models can be used if one fails a new one can be trained leading to an approach called continuous verification, referring to the loop between training and verification. One tactic for improving the network's performance is through "constraint-based loss functions" - a method of using differentiable logic (DL) to translate logical constraints into loss functions which can then be used to train the network specifically to satisfy said constraint. In this paper we present a uniform way of defining a translation from logic syntax to a differentiable loss function then examine and compare the existing DLs. We explore mathematical properties desired in such translations and discuss the design space identifying possible directions of future work.

1   Introduction

The rising popularity of neural networks (NNs) in recent years and their increasing prevalence in real-world applications have drawn attention to the importance of their verification. While verification is known to be computationally difficult theoretically [6], many techniques have been proposed for solving it in practice [1].

It has been observed in the literature that by default neural networks rarely satisfy logical constraints that we want to verify. A good course of action is to train the given NN to satisfy said constraint prior to verifying them [5,15]. This idea is sometimes referred to as continuous verification [2,9], referring to the loop between training and verification.

Usually training with constraints is implemented by specifying a translation for a given formal logic language into loss functions. These loss functions are then used to train neural networks. Because for training purposes these functions need to be differentiable, these translations are called *differentiable logics* (DL).

This raises several research questions. What kind of differentiable logics are possible? What difference does a specific choice of DL make in the context of

© The Author(s), under exclusive license to Springer Nature Switzerland AG 2022
O. Isac et al. (Eds.): NSV 2022/FoMLAS 2022, LNCS 13466, pp. 67–77, 2022.
https://doi.org/10.1007/978-3-031-21222-2_5

continuous verification? What are the desirable criteria for a DL viewed from the point of view of the resulting loss function? In this extended abstract we will discuss and answer these questions.

2   Differentiable Logic and Constraint Cased Loss Functions

We will explain the main idea behind DL by means of concrete examples. Recall that in machine learning loss functions are used during training to calculate the error between the neural network's current output and the desired output. For example cross-entropy, a popular choice of loss function for classification problems, is defined as follows:

Definition 1 (Cross Entropy Loss Function). *Given a data set* $\mathcal{D} = \{(\mathbf{x}_1, \mathbf{y}_1), \ldots, (\mathbf{x}_n, \mathbf{y}_n)\}$ *and a function (neural network)* $f : \mathbb{R}^n \to [0,1]^m$, *the cross-entropy loss is defined as*

$$\mathcal{L}_{ce}(\mathbf{x}, \mathbf{y}) = -\sum_{i=1}^{m} \mathbf{y}_i \, \log(f(\mathbf{x})_i) \tag{1}$$

where $(\mathbf{x}, \mathbf{y}) \in \mathcal{D}$, $\mathbf{y}_i$ *is the true probability for class* i *and* $f(\mathbf{x})_i$ *the probability for class* i *as predicted by* f *when applied to* $\mathbf{x}$.

We now define a small formal language Φ: let the terms $t, t' \in T$ be either variables or constants. Atomic propositions (also called atomic formulae) $A, A_i \in \mathcal{A}$ are given by comparisons between terms using binary predicates $\leq, \neq$. Let us also have conjunction $\wedge$ and negation $\neg$.

$$\Phi \ni \phi, \phi_1, \phi_2 = A \mid \phi_1 \wedge \phi_2 \mid \neg\phi$$

Let us use this toy syntax to introduce several different DLs that exist in the literature. To do so we will use the notation $\| \cdot \| : \Phi \to \mathcal{D}$ for some target domain $\mathcal{D}$ to talk about the possible translations.

DL2 Translation. A good first example is a system called DL2 (Deep Learning with Differentiable Logic) [4]. The translation function $\| \cdot \|_{DL2} : \Phi \to [0, \infty]$ is defined as follows.

The definition starts with atomic formulae. The loss for $t \leq t'$ is proportional to the difference between terms t and t' and the loss for term inequality is non-zero when the terms are equal:

$$\|t \leq t'\|_{DL2} := \max(t - t', 0)$$
$$\|t \neq t'\|_{DL2} := \xi[t = t']$$

where $\xi > 0$ is a constant and $[\cdot]$ an indicator function. And conjunction is defined as:

$$\|\phi_1 \wedge \phi_2\|_{DL2} := \|\phi_1\|_{DL2} + \|\phi_2\|_{DL2} \tag{2}$$

We notice that negation is not explicitly defined in DL2 - it is only defined for atomic formulae since they are comparisons. This is partially related to the choice of domain of the function $\| \cdot \|_{DL2}$. In the interval $[0, \infty]$, 0 denotes **true** and the rest of the interval denotes **false**. This interpretation does not admit a simple operation for inversion that could give an interpretation for negation. More generally the choice of interpretation range $[0, \infty]$ is motivated by making the resulting function differentiable "almost everywhere" (this range resembles the famous activation function ReLu) and give an interpretation of the basic predicates $\leq, \neq$.

Note that we can also view the above translation as a function. For example:

$$\| \leq \|_{DL2} = \lambda t, t'. \ max(t - t', 0)$$
$$\| \neq \|_{DL2} = \lambda t, t'. \ \xi[t = t']$$
$$\| \leq \|_{DL2} = \lambda \phi_1, \phi_2. \ \|\phi_1\|_{DL2} + \|\phi_2\|_{DL2}$$

In the later sections we will sometimes resort to the functional notation for the translation function.

Fuzzy Logic Translation. Fuzzy DL takes a more conceptual approach to the choice of the domain $\mathcal{D}$. Unlike DL2 which had a focus on interpreting comparisons between terms, fuzzy DL starts with the domain intrinsic to fuzzy logic and develops the DL translation from there [10–12].

Using our example we look at one implementation of conjunction in fuzzy logic based on Gödel's t-norm (t-norm, or a triangular norm, is a kind of binary operation used, among others, in fuzzy logic [3]). Let us denote this translation as $\| \cdot \|_G : \Phi \rightarrow [0, 1]$. This time for the base case we assume that atomic formulae are mapped to $[0, 1]$ by some oracle. Then:

$$\|\phi_1 \wedge \phi_2\|_G := \min(\|\phi_1\|_G, \|\phi_2\|_G) \tag{3}$$

This time we are given a translation for negation which is

$$\|\neg\phi\|_G := 1 - \|\phi\|_G$$

Drawing from the long tradition of fuzzy logic research, van Krieken et al. [12] survey several other choices of translation for conjunction in fuzzy logic such as Łukasiewicz:

$$\|\phi_1 \wedge \phi_2\|_L = \max(\|\phi_1\|_L + \|\phi_2\|_L - 1, 0))$$

Yager:

$$\|\phi_1 \wedge \phi_2\|_Y = \max(1 - ((1 - \|\phi_1\|_Y)^p + (1 - \|\phi_2\|_Y)^p)^{1/p}, 0), \ \text{ for } p \geq 1$$

or product based:

$$\|\phi_1 \wedge \phi_2\|_P = \|\phi_1\|_P \cdot \|\phi_2\|_P$$

to name a few (see [12] for full survey). All of these logics agree on interpretation of negation.

Signal Temporal Logic Translation. A different approach by Varnai and Dimarogonas [13] was suggested for Signal Temporal Logic (STL) which we adapt to our example language. This paper suggests that the design of DLs should focus on specific properties of the loss functions they give rise to.

Let us denote the new translation $\|\cdot\|_S : \Phi \to \mathbb{R}$. Varnai and Dimarogonas [13] start with a list of desirable properties of loss functions and create the translation around it. In all of the following we will use $\bigwedge_M(A_1, ..., A_M)$ as a notation for conjunction of exactly M conjuncts. Again the authors assume an oracle (map of atomic formulae to $\mathbb{R}$) for the base case.

We assume a constant $\nu \in \mathbb{R}^+$ and we introduce the following notation:

$$A_{\min} = \min(\|A_1\|_S, \ldots, \|A_M\|_S)$$

and

$$\tilde{A}_i = \frac{\|A_i\|_S - A_{\min}}{A_{\min}}$$

Then the translation is defined as:

$$\left\| \bigwedge_M(A_1, \ldots, A_M) \right\|_S = \begin{cases} \dfrac{\sum_i A_{\min} e^{\tilde{A}_i} e^{\nu \tilde{A}_i}}{\sum_i e^{\nu \tilde{A}_i}} & \text{if } A_{\min} < 0 \\ \dfrac{\sum_i A_{\min} e^{-\nu \tilde{A}_i}}{\sum_i e^{-\nu \tilde{A}_i}} & \text{if } A_{\min} > 0 \\ 0 & \text{if } A_{\min} = 0 \end{cases} \tag{4}$$

This translation proposes an elegant notion of negation:

$$\|\neg\phi\|_S = -\|\phi\|_S$$

Use in Training. To use any of these functions in NN training we augment the standard loss function. For example given cross-entropy loss (see Definition 1), a constraint ϕ and a translation $\|\cdot\|$ we would have an augmented loss function:

$$\mathcal{L}_A^\phi(\mathbf{x}, \mathbf{y}) = \alpha \mathcal{L}_{CE}(\mathbf{x}, \mathbf{y}) + \beta\|\phi\| \tag{5}$$

where $\alpha, \beta \in [0, 1]$ are constants.

By looking at these key choices made in literature so far we can see that the choice of a DL involves four major decisions:

- **Domain of interpretation.** We have seen the choices vary between $[0, \infty]$, $[0, 1]$ and $[-\infty, +\infty]$.
- **Expressiveness.** Choice of which connectives will be included in the translation, which determines the expressiveness of the language.
- **Interpreting logical connectives.** Although negation is partially determined by the choice of domain, the choice of conjunction seems to be a largely independent decision as evidenced by the presented examples.
- **Interpretation of atomic formulae.** DL2 proposes a concrete approach and some papers leave the definition abstract.

In context of continuous verification these choices determine how the translation is implemented in practice.

3 Property Based Approach

We now consider the mathematical properties of the resulting loss functions. Generally, machine learning research suggests a choice of loss functions for NNs: cross-entropy loss, hinge loss, mean squared error loss, etc. [14]. In this community there is some consensus on the desirable properties of loss functions - convexity or continuity are widely considered desirable [7]. But as we are now focusing on developing methods for logic-driven loss functions, Varnai and Dimarogonas [13] also suggest additional desirable properties which we will discuss next. Following from the previous section we continue to assume an oracle mapping of all atomic formulae $\| \cdot \| : \mathcal{A} \to \mathcal{D}$.

Soundness relates to the logical satisfaction of the formula. We assume that the language Φ has some interpretation $\mathcal{I}$ of its formulae to the set {**true, false**}. As discussed in Sect. 2, the domain in each of the translations is different and that heavily influences what *soundness* will be defined as. Given a domain $\mathcal{D}$ we must divide it into the parts that map to **true** and **false**. Let us denote the part that maps to **true** as $\mathcal{D}_{true}$.

Let us now define the soundness abstractly.

Definition 2 (Soundness). *A DL is sound for some interpretation $\mathcal{I}$ if for any constraint ϕ, $\|\phi\| \in \mathcal{D}_{true}$ if and only if constraint ϕ is true in interpretation $\mathcal{I}$, denoted as $\mathcal{I}(\phi) = \mathbf{true}$.*

Let us compare the soundness for the specific translations starting from DL2: $\| \cdot \|_{DL2} : \Phi \to [0, \infty]$:

$$\mathcal{I}(\phi) = \mathbf{true} \Leftrightarrow \|\phi\|_{DL2} = 0 \tag{6}$$

For the fuzzy logic we have: $\| \cdot \|_G : \Phi \to [0, 1]$. This is a more intriguing problem as we can no longer assume ϕ is interpreted on boolean values. There is now a choice of splitting the domain to adhere to boolean values as we have used above or using fuzzy logic to express the constraints. In this extended abstract we will use the absolute truth from fuzzy logic instead:

$$\mathcal{I}(\phi) = 1 \Leftrightarrow \|\phi\|_G = 1 \tag{7}$$

And lastly for the STL translation $\| \cdot \|_S : \Phi \to \mathbb{R}$, we have:

$$\mathcal{I}(\phi) = \mathbf{true} \Leftrightarrow \|\phi\|_S > 0 \tag{8}$$

As we can see the soundness here is different from how loss functions are usually constructed – they usually have values greater or equal to zero – this choice however is connected to the shadow-lifting property which we discuss later.

These are all significantly different soundness statements which poses a question of how much does the choice of logic impact the mathematical properties of the function that we get from the translation – or conversely whether some properties we may want our function to have can limit the choice of logic we can use.

Definition 3 (Commutativity, Idempotence and Associativity). *The* ***and*** *operator* $\bigwedge_M$ *is commutative if for any permutation* π *of the integers* $i \in 1, ..., M$

$$\| \bigwedge_M (A_1, ..., A_M) \| = \| \bigwedge_M (A_{k_{\pi(1)}}, ..., A_{k_{\pi(M)}}) \|$$

it is idempotent if

$$\| \bigwedge_M (A, ..., A) \| = \|A\|$$

and associative if

$$\| \bigwedge_2 (\bigwedge_2 (A_1, A_2), A_3) \| = \| \bigwedge_2 (A_1, \bigwedge_2 (A_2, A_3)) \|$$

Commutativity, idempotence and associativity are identities that make it far easier to use the translation, as changes in the order of elements in conjunction will not affect the resulting loss function. It is also important to note that associativity is not a part of original set of desirable properties as listed by Varnai and Dimarogonas [13] and is not satisfied by the translation $\| \cdot \|_S$ – which is the reason why they define conjunction as an n-ary rather than a binary operator.

Before defining the next property we should also take a look at the notion of *gradient*, which for a differentiable function f is a vector of its partial derivatives at a given point.

Definition 4 (Weak smoothness). *The* $\| \bigwedge_M \|$ *is weakly smooth if it is continuous everywhere and its gradient is continuous at all points* $\|A_i\|$ *with* $i \in \{1, ..., M\}$ *for which no two indices* $i \neq j$ *satisfy* $\|A_i\| = \|A_j\| = \min(\|A_1\|, ..., \|A_M\|)$.

In more informal terms we require smoothness at points where there is a unique minimal term. The specific definition of weak-smoothness holds in particular for points where the metric switches signs (and therefore, by definition, its truth value).

Definition 5 (Min-max boundedness). *The operator* $\| \bigwedge_M \|$ *is min-max bounded if*

$$\min(\|A_1\|, ..., \|A_M\|) \leq \| \bigwedge_M (A_1, ..., A_M) \| \leq \max(\|A_1\|, ..., \|A_M\|)$$

The min-max boundedness ensures closure of the translation with respect to the domain.

Definition 6 (Scale invariance). *The interpretation of* $\bigwedge_M$ *is said to be scale-invariant if, for any* $\alpha \leq 0$ *with* $\alpha \in \mathbb{R}$

$$\alpha \| \bigwedge_M (A_1, ..., A_M) \| = \left\| \bigwedge_M \right\| (\alpha \|A_1\|, ..., \alpha \|A_M\|)$$

With **scale-invariance** we can be sure that the metric will behave in a similar manner regardless of the magnitude (in case it was unknown).

Now "shadow-lifting" is a property original to [13]. Its motivation is to encourage gradual improvement when training the neural network even when the property is not yet satisfied. In other words if one part of the conjunction increases the value for the translation of entire conjunction should increase as well.

Definition 7 (Shadow-lifting property). *Let* $\| \bigwedge_M \|$ *satisfy the shadow-lifting property if,* $\forall i. \|A_i\| \neq 0$:

$$\frac{\partial \| \bigwedge_M (A_1, ..., A_M) \|}{\partial \|A_i\|} \Bigg|_{A_1, ..., A_M} > 0$$

where ∂ *denotes partial differentiation.*

It is this property that motivated the translation of conjunction $\| \cdot \|_S$ described in Eq. 4. It is different to the classical notion of conjunction which, if not satisfied, does not preserve information when it comes to value of individual conjuncts or their number. It is interesting to consider whether it would be more useful to use a logic with two variants of each connective – keeping the classical FOL connectives as well as ones adhering to the shadow-lifting property.

There is also the issue of domain which we've touched on before. This translation allows the values of the translation to range across entire real domain ($\mathbb{R}$) – defining *true* as greater then zero and allowing us to define negation by simply flipping the sign. While this approach was viable for reinforcement learning that this metric was designed for, it creates a problem of compatibility when it comes to training neural networks. Typically a constraint loss function that is generated is not used on its own – but in combination with a more classical one such as cross-entropy loss (see Eq. 5)– and would have to be scaled appropriately if we do not want it to imbalance the training.

This poses a more general question of how the choice of properties that we deem as desirable determines the choice of logical syntax that we can use.

Let us see how the DLs that we have introduced compare when it comes to satisfying these properties.

Table 1 summarises results already established in literature as well as provides new results. Let us briefly describe the reasoning behind the table entries that gave new results. It is important to note that the properties take into account the domains of each translation.

Starting with shadow-lifting, Gödel, Łukasiewicz and Yager t-norms all do not have that property. These translations involve minimum or maximum which both do not preserve shadow-lifting. In all of these cases there are cases at which change of value of one of the conjuncts will not change the value of min\max. But the property holds for both DL2 and product translations for most properties due to them being defined simply by addition and multiplication respectively.

With min-max boundedness the reasoning is different for each translation. Interestingly here the product translation is bounded due to the domain being $[0, 1]$. By definition of minimum the Gödel based DL is also bounded while both

Table 1. Property comparison between different DL translations of conjunction. Properties which have been stated in the relevant paper or proven in other works have relevant citations.

	$\|\phi_1 \wedge \phi_2\|_{DL2}$	$\|\phi_1 \wedge \phi_2\|_G$	$\|\phi_1 \wedge \phi_2\|_L$	$\|\phi_1 \wedge \phi_2\|_Y$	$\|\phi_1 \wedge \phi_2\|_P$	$\|\phi_1 \wedge \phi_2\|_S$
	$[0, \infty]$	$[0, 1]$	$[0, 1]$	$[0, 1]$	$[0, 1]$	$[-\infty, \infty]$
Properties:						
Idempotent	No	Yes [12]	No [3]	no [8]	no [3]	yes [13]
Commutative	Yes	Yes [12]	Yes [12]	yes [12]	yes [12]	yes [13]
Shadow-lifting	Yes	No	No	no	yes	yes [13]
Min-max boundedness	No	Yes	No	no	yes	yes [13]
Scale invariance	Yes	Yes	No	no	no	yes [13]
Associativity	Yes	Yes [12]	yes [12]	yes [12]	yes [12]	no [13]

Łukasiewicz and Yager can return values greater then the largest value of a conjunct.

For the case of scale invariance DL2, Gödel and product entries are trivial. Both Łukasiewicz and Yager inspired DLs are not scale invariant due to the terms inside max containing addition or subtraction of constants not dependant on the individual conjuncts.

All of the fuzzy logic translations are associative as they are based on t-norms which are associative by definition [12] and DL2 is associative as it is defined by addition. The STL based metric is the only one which is not associative – associativity together with idempotence would prevent shadow-lifting which was deemed more desirable [13].

We can see that none of the translations have all of these properties. This shows that the property oriented approach to finding a translation is non-trivial and the choice of a DL heavily influences the properties of the resulting loss function.

4   Conclusions and Future Work

4.1   Conclusions

In this extended abstract we answered the questions posed in the abstract. Firstly, we presented a uniform way of defining a translation from logic syntax to a differentiable loss function. This has allowed us to compare them in terms of the mathematical properties of their translation, providing an overview of the current state of the art when it comes to DLs. This in turn allowed us to reason about the design space of future DLs – the properties we may want them to have, the choice of domain, choice of logic etc. This is the first step to providing a comprehensive guide to creating new DLs while being conscious of the consequences certain design decisions can bring for continuous verification of NNs.

4.2   Future Work and Design Space

We have briefly mentioned at the end of Sect. 2 some of the design choices that one has to face when choosing and designing a DL, as well as its interpretation to a loss function for training neural networks. When we try to compare the different DLs we can group the trade-offs in a few categories:

Expressive Power of DLs. *Expressive power* is a broad category that we have mentioned in Sect. 2. It is impacted by the choice of logic and its domain of interpretation as those two things can limit, for example, the choice of defined connectives as can be seen in the following example of DL2.

The lack of negation in the defined syntax of DL2 is a direct consequence of the domain ($\mathcal{D} = [0, \infty]$). The only reason the translation does not lead to a significant loss of expressiveness is due to the explicit translation of all atomic formulae and their negations – this way one can "push" the negation inwards in any constraint. In case all other translations discussed before, which all leave the interpretation of atomic formulae to an oracle, it would only work if we added an assumption that said oracle also interprets negation of every atomic formula.

Meanwhile for fuzzy logic DLs there is a well defined domain, however we encounter an issue when trying to split it to assign boolean values for the purposes of translation. We need to have a fully **true** state – state when the constraint is fully satisfied. This creates a choice of how the domain should be split, which heavily impacts the expressiveness of the DL.

Mathematical Properties of DLs. We have already discussed how different DLs compare in terms of *mathematical properties* (see Sect. 3). Some of these properties are matters of convenience – associativity and commutativity for example ensure that order of elements in conjunction will not affect the resulting loss function. Others, such as shadow-lifting, change the way the loss function will penalise NNs behaviour and therefore the way it is trained.

This comparison comes down to interpretation of logical connectives. We have presented multiple interpretations of conjunction which, while of course influenced by the domain, give a lot of freedom in their design. Fuzzy logic based DLs are a great example of this considering they resulted in multiple translations with the same domain and syntax and different mathematical properties and behaviour (see Sect. 3 for a detailed discussion).

While both the syntax being translated and the mapping of values are dependant on properties the properties themselves are not immutable. This leads to a question of what other properties would we want to add to the list of ones we consider desirable and which should be prioritised – for example among the current properties it is impossible to satisfy idempotence, associativity and shadow-lifting simultaneously. It is also important to mention that in this study we have we omitted completeness in this study, yet it deserves further investigation in the future.

Training Efficacy and Efficiency. From an implementation perspective there is also the question of *efficiency* which we have not discussed before – for example we may prefer to avoid non-transcendental functions or case splits – both of which were present in some translations – as they make it far more costly to train the network.

An immediate plan for the future involves developing a new translation into a differentiable loss function, taking some inspiration from the property-driven approach. In this talk we will discuss how the different translations presented compare when it comes to performance – based on the results we will be able to draw more conclusions about the design space for DLs in future work.

Acknowledgement. Authors acknowledge support of EPSRC grant AISEC EP/T026952/1 and NCSC grant Neural Network Verification: in search of the missing spec.

References

1. Bak, S., Liu, C., Johnson, T.: The second international verification of neural networks competition (VNN-COMP 2021): summary and results, 2021. Technical Report. https://arxiv.org/abs/2109.00498
2. Casadio, M., et al.: Neural network robustness as a verification property: a principled case study. In: Computer Aided Verification (CAV 2022), Lecture Notes in Computer Science. Springer, 2022 https://doi.org/10.1007/978-3-031-13185-1_11
3. Cintula, P., Hájek, P., Noguera, C.: Handbook of mathematical fuzzy logic (in 2 volumes), vol. 37, 38 of studies in logic, mathematical logic and foundations (2011)
4. Fischer, M., Balunovic, M., Drachsler-Cohen, D., Gehr, T., Zhang, C., Vechev, M.: DL2: training and querying neural networks with logic (2019). https://openreview.net/forum?id=H1faSn0qY7
5. Hu, Z., Ma, X., Liu, Z., Hovy, E., Xing, E.: Harnessing deep neural networks with logic rules. arXiv preprint arXiv:1603.06318 (2016)
6. Katz, G., Barrett, C., Dill, D.L., Julian, K., Kochenderfer, M.J.: Reluplex: an efficient SMT solver for verifying deep neural networks. In: Majumdar, R., Kunčak, V. (eds.) CAV 2017. LNCS, vol. 10426, pp. 97–117. Springer, Cham (2017). https://doi.org/10.1007/978-3-319-63387-9_5
7. Klebanov, L.B., Rachev, S.T., Fabozzi, F.J.: Robust and non-robust models in statistics. Nova Science Publishers Hauppauge (2009)
8. Klement, E.P., Mesiar, R., Pap, E.: Triangular norms. position paper ii: general constructions and parameterized families. Fuzzy sets Syst. **145** (3), 411–438 (2004)
9. Komendantskaya, E., Kokke, W., Kienitz, D.: Continuous verification of machine learning: a declarative programming approach. In: PPDP 2020: 22nd International Symposium on Principles and Practice of Declarative Programming, Bologna, Italy, pp. 1:1–1:3. ACM (2020)
10. Serafini, L., d'Avila Garcez, A.S.: Logic tensor networks: deep learning and logical reasoning from data and knowledge. CoRR abs/1606.04422 (2016)
11. Krieken, E.V., Acar, E., Harmelen, F.V.: Analyzing differentiable fuzzy implications. CoRR abs/2006.03472 (2020)
12. Krieken, E.V., Acar, E., Harmelen, F.V.: Analyzing differentiable fuzzy logic operators. Artif. Intell. **302**, 1–46 (2022)

13. Varnai, P., Dimarogonas, D.: On robustness metrics for learning STL tasks (2020)
14. Wang, Q., Ma, Y., Zhao, K., Tian, Y.: A comprehensive survey of loss functions in machine learning. Ann. Data. Sci. **9**(2), 187–212 (2022)
15. Xu, J., Zhang, Z., Friedman, T., Liang, Y., Van den Broeck, G.: a semantic loss function for deep learning with symbolic knowledge. In: Dy, J., Krause, A., eds, Proceedings of the 35th International Conference on Machine Learning, vol. 80 of Proceedings of Machine Learning Research, pp. 5502–5511. PMLR, 10–15 Jul 2018. https://proceedings.mlr.press/v80/xu18h.html

Neural Networks in Imandra: Matrix Representation as a Verification Choice

Remi Desmartin[1]([⊠]), Grant Passmore[2], and Ekaterina Kommendentskaya[1]

[1] Heriot -Watt University, Edinburgh, UK
{rhd2000,e.komendantskaya}@hw.ac.uk
[2] Imandra, Austin, TX, USA
grant@imandra.ai
http://www.imandra.ai

Abstract. The demand for formal verification tools for neural networks has increased as neural networks have been deployed in a growing number of safety-critical applications. Matrices are a data structure essential to formalising neural networks. Functional programming languages encourage diverse approaches to matrix definitions. This feature has already been successfully exploited in different applications. The question we ask is whether, and how, these ideas can be applied in neural network verification. A functional programming language Imandra combines the syntax of a functional programming language and the power of an automated theorem prover. Using these two key features of Imandra, we explore how different implementations of matrices can influence automation of neural network verification.

Keywords: Neural networks · Matrices · Formal verification · Functional programming · Imandra

1 Motivation

Neural network (NN) verification was pioneered by the SMT-solving [11,12] and an abstract interpretation [2,8,20] communities. However, recently claims have been made that functional programming, too, can be valuable in this domain. There is a library [16] formalising small rational-valued neural networks in Coq. A more sizeable formalisation called MLCert [3] imports neural networks from Python, treats floating point numbers as bit vectors, and proves properties describing the generalisation bounds for the neural networks. An F^* formalisation [14] uses F^* reals and refinement types for proving the robustness of networks trained in Python.

There are several options for defining neural networks in functional programming, ranging from defining neurons as record types [16] to treating them as functions with refinement types [14]. But we claim that two general considerations should be key to any NN formalisation choice. Firstly, we must define neural networks as executable functions, because we want to take advantage of

E. Komendantskaya—Acknowledges support of EPSRC grant EP/T026952/1.

O. Isac et al. (Eds.): NSV 2022/FoMLAS 2022, LNCS 13466, pp. 78–95, 2022.
https://doi.org/10.1007/978-3-031-21222-2_6

executing them in the functional language of choice. Secondly, a generic approach to layer definitions is needed, particularly when we implement complex neural network architectures, such as convolutional layers.

These two essential requirements dictate that neural networks are represented as matrices, and that a programmer makes choices about matrix formalisation. This article will explain these choices, and the consequences they imply, from the verification point of view. We use Imandra [17] to make these points, because Imandra is a functional programming language with tight integration of automated proving.

Imandra has been successful as a user-friendly and scalable tool in the FinTech domain [18]. The secret of its success lies in a combination of many of the best features of functional languages and interactive and automated theorem provers. Imandra's logic is based on a pure, higher-order subset of OCaml, and functions written in Imandra are at the same time valid OCaml code that can be executed, or *"simulated"*. Imandra's mode of interactive proof development is based on a typed, higher-order lifting of the *Boyer-Moore waterfall* [4] for automated induction, tightly integrated with novel techniques for SMT modulo recursive functions.

This paper builds upon the recent development of a CheckINN, a NN verification library in Imandra [6], but discusses specifically the matrix representation choices and their consequences.

2   Matrices in Neural Network Formalisation

We will illustrate the functional approach to neural network formalisation and will introduce the syntax of the Imandra programming language [17] by means of an example. When we say we want to formalise neural networks as functions, essentially, we aim to be able to define a NN using just a line of code:

```
let cnn input =
    layer_0 input >>= layer_1 >>= layer_2 >>= layer_3
```

where each `layer_i` is defined in a modular fashion.

To see that a functional approach to neural networks does not necessarily imply generic nature of the code, let us consider an example. A *perceptron*, also known as a *linear classifier*, classifies a given input vector $X = (x_1, ..., x_m)$ into one of two classes c_1 or c_2 by computing a linear combination of the input vector with a vector of synaptic weights $(w_0, w_1, ..., w_m)$, in which w_0 is often called an *intercept* or *bias*: $f(X) = \sum_{i=1}^{m} w_i x_i + w_0$. If the result is positive, it classifies the input as c_1 and if negative as c_2. It effectively divides the input space along a hyperplane defined by $\sum_{i=1}^{m} w_i x_i + w_0 = 0$.

In most classification problems, classes are not linearly separated. To handle such problems, we can apply a non-linear function a called an *activation function* to the linear combination of weights and inputs. The resulting definition of a perceptron f is:

$$f(X) = a \left(\sum_{i=1}^{m} w_i x_i + w_0 \right) \tag{1}$$

Let us start with a naive prototype of perceptron in Imandra. The Iris data set is a "Hello World" example in data mining; it represents 3 kinds of Iris flowers using 4 selected features. In Imandra, inputs can be represented as a data type:

```
type iris_input = {
  sepal_len: real;
  sepal_width: real;
  petal_len: real;
  petal_width: real;}
```

And we define a perceptron as a function:

```
let layer_0 (w0, w1, w2, w3, w4) (x1, x2, x3, x4) =
  relu (w0 +. w1 *. x1 +. w2 *. x2 +. w3 *. x3 +. w4 *. x4)
```

where *. and +. are *times* and *plus* defined on reals. Note the use of the `relu` activation function, which returns 0 for all negative inputs and acts as the identity function otherwise.

Already in this simple example, one perceptron is not sufficient, as we must map its output to three classes. We use the usual machine learning literature trick and define a further layer of 3 neurons, each representing one class. Each of these neurons is itself a perceptron, with one incoming weight and one bias. This gives us:

```
let layer_1 (w1, b1, w2, b2, w3, b3) f1 =
  let o1 = w1 *. f1 +. b1 in
  let o2 = w2 *. f1 +. b2 in
  let o3 = w3 *. f1 +. b3 in
  (o1, o2, o3)

let process_iris_output (c0, c1, c2) =
  if (c0 >=. c1) && (c0 >=. c2) then "setosa"
  else if (c1 >=. c0) && (c1 >=. c2) then "versicolor"
  else "virginica"
```

The second function maps the output of the three neurons to the three specified classes. This post-processing stage often takes the form of an *argmax* or *softmax* function, which we omit.

And thus the resulting function that defines our neural network model is:

```
let model input = process_iris_input input
        |> layer_0 weights_0 |> layer_1 weights_1 |>
                            process_iris_output
```

Although our naive formalisation has some features that we desired from the start, i.e. it defines a neural network as a composition of functions, it is too inflexible to work with arbitrary compositions of layers. In neural networks with hundreds of weights in every layer this manual approach will quickly become infeasible (as well as error-prone). So, let us generalise this attempt from the level of individual neurons to the level of matrix operations.

The composition of many perceptrons is often called a *multi-layer perceptron (MLP)*. An MLP consists of an input vector (also called input layer in the literature), multiple hidden layers and an output layer, each layer made of perceptrons with weighted connections to the previous layers' outputs. The weight and biases of all the neurons in a layer can be represented by two matrices denoted by W and B. By adapting Eq. 1 to this matrix notation, a layer's output L can be defined as:

$$L(X) = a(X \cdot W + B) \qquad (2)$$

where the operator $\cdot$ denotes the dot product between X and each row of W, X is the layer's input and a is the activation function shared by all nodes in a layer. As the dot product multiplies pointwise all inputs by all weights, such layers are often called *fully connected*.

By denoting a_k, W_k, B_k — the activation function, weights and biases of the kth layer respectively, an MLP F with L layers is traditionally defined as:

$$F(X) = a_L[B_L + W_L(a_{L-1}(B_{L-1} + W_{L-1}(...(a_1(B_1 + W_1 \cdot X)))))] \qquad (3)$$

At this stage, we are firmly committed to using matrices and matrix operations. And we have two key choices:

1. to represent matrices as lists of lists (and take advantage of the inductive data type `List`),
2. define matrices as functions from indices to matrix elements,
3. or take advantage of record types, and define matrices as records with maps.

The first choice was taken in [10] (in the context of dependent types in Coq), in [14] (in the context of refinement types of F*) and in [9] (for sparse matrix encodings in Haskell). The difference between the first and second approaches was discussed in [22] (in Agda, but with no neural network application in mind). The third method was taken in [16] using Coq (records were used there to encode individual neurons).

In the next three sections, we will systematise these three approaches using the same formalism and the same language, and trace the influence of these choices on neural network verification.

3 Matrices as Lists of Lists

We start with re-using Imandra's `List` library. Lists are defined as inductive data structures:

```
type 'a list =
| []
| (::) of 'a * 'a list
```

Imandra holds a comprehensive library of list operations covering a large part of OCaml's standard `List` libary, which we re-use in the definitions below. We start with defining vectors as lists, and matrices as lists of vectors.

```
type 'a vector = 'a list
type 'a matrix = 'a vector list
```

It is possible to extend this formalisation by using dependent [10] or refinement [14] types to check the matrix size, e.g. when performing matrix multiplication. But in Imandra this facility is not directly available, and we will need to use error-tracking (implemented via the monadic `Result` type) to facilitate checking of the matrix sizes.

As there is no built-in type available for matrices equivalent to `List` for vectors, the `Matrix` module implements a number of functions for basic operations needed throughout the implementation. For instance, `map2` takes as inputs a function f and two matrices A and B of the same dimensions and outputs a new matrix C where each element $c_{i,j}$ is the result of $f(a_{i,j}, b_{i,j})$:

```
let rec map2 (f:'a -> 'b -> 'c) (x:'a matrix) (y:'b matrix)
    =
    match x with
    | [] -> begin match y with
            | [] -> Ok []
            | y::ys  -> Error "map2: invalid length." end
    | x::xs -> begin match y with
               | [] -> Error "map2: invalid length."
               | y::ys -> let hd = map2 f x y in
                          let tl = map2 f xs ys in
                          lift2 cons hd tl end
```

This implementation allows us to define other useful functions concisely. For instance, the dot-product of two matrices is defined as:

```
let dot_product (a:real matrix) (b:real matrix): ('a, real
    matrix) result =
  Result.map sum (map2 ( *. ) a b)
```

Note that since the output of the function `map2` is wrapped in the monadic `result` type, we must use `Result.map` to apply `sum`. Similarly, we use standard monadic operations for the `result` monad such as `bind` or `lift`.

A fully connected layer is then defined as a function `fc` that takes as parameters an activation function, a 2-dimensional matrix of layer's weights and an input vector:

```
let activation f w i = (* activation func., weights, input *)
  let linear_combination m1 m2 = if (length m1) <> (length m2)
```

```
      then Error "invalid dimensions"
      else map sum (Vec.map2 ( *. ) m1 m2) in
 let i' = 1.::i in (* prepend 1. for bias *)
 let z = linear_combination w i' in
 map f z

let rec fc f (weights:real matrix) (input:real vector) =
 match weights with
 | [] -> Ok []
 | w::ws -> lift2 cons (activation f w input) (fc f ws input)
```

Listing 1.1. Fully connected layer implementation

Note that each row of the weights matrix represents the weights for one of the layer's nodes. The bias for each node is the first value of the weights vector, and 1 is prepended to the input vector when computing the dot-product of weights and input to account for that.

It is now easy to see that our desired modular approach to composing layers works as stated. We may define the layers using the syntax: `let layer_i = fc a weights`, where i stands for 0,1,2,3, and a stands for any chosen activation function.

Although natural, this formalisation of layers and networks suffers from two problems. Firstly, it lacks the matrix dimension checks that were readily provided via refinement types in [14]. This is because Imandra is based on a computational fragment of HOL, and has no refinement or dependent types. To mitigate this, our library performs explicit dimension checking via a `result` monad, which clutters the code and adds additional computational checks. Secondly, the matrix definition via the list datatypes makes verification of neural networks very inefficient. This general effect has been already reported in [14], but it may be instructive to look into the problem from the Imandra perspective.

Robustness of neural networks [5] is best amenable to proofs by arithmetic manipulation. This explains the interest of the SMT-solving community in the topic, which started with using Z3 directly [11], and has resulted in highly efficient SMT solvers specialised on robustness proofs for neural networks [12,13]. Imandra's waterfall method [17] defines a default flow for the proof search, which starts with unrolling inductive definitions, simplification and rewriting. As a result, proofs of neural network robustness or proofs as in the ACAS Xu challenge [12,13], which do not actually need induction, are not efficiently tackled using Imandra's inductive waterfall: the proofs simply do not terminate.

There is another verification approach available in Imandra which is better suited for this type of problem: `blast`, a tactic for SAT-based symbolic execution modulo higher-order recursive functions. Blast is an internal custom SAT/SMT solver that can be called explicitly to discharge an Imandra verification goal. However, `blast` currently does not support real arithmetic. This requires us to *quantize* the neural networks we use (i.e. convert them to integer weights) and results in a *quantised NN implementation* [6]. However, even with quantisation and the use of `blast`, while we succeed on many smaller benchmarks, Imandra fails to scale 'out of the box' to the ACAS Xu challenge, let alone larger neural networks used in computer vision.

This also does not come as a surprise: as [12] points out, general-purpose SMT solvers do not scale to NN verification challenges. This is why the algorithm `reluplex` was introduced in [12] as an additional heuristic to SMT solver algorithms; `reluplex` has since given rise to a domain specific solver Marabou [13]. Connecting Imandra to Marabou may be a promising future direction.

However, this method of matrix formalisation can still bring benefits. When we formulate verification properties that genuinely require induction, formalisation of matrices as lists does result in more natural, and easily automatable proofs. For example, De Maria et al. [16] formalise in Coq *"neuronal archetypes"* for biological neurons. Each archetype is a specialised kind of perceptron, in which additional functions are added to amplify or inhibit the perceptron's outputs. It is out of the scope of this paper to formalise the neuronal archetypes in Imandra, but we take methodological insight from [16]. In particular, De Maria et al. show that there are natural higher-order properties that one may want to verify.

To make a direct comparison, modern neural network verifiers [12,20] deal with verification tasks of the form "given a trained neural network f, and a property P_1 on its inputs, verify that a property P_2 holds for f's outputs". However, the formalisation in [16] considers properties of the form "any neural network f that satisfies a property Q_1, also satisfies a property Q_2." Unsurprisingly, the former kind of properties can be resolved by simplification and arithmetic, whereas the latter kind requires induction on the structure of f (as well as possibly nested induction on parameters of Q_1).

Another distinguishing consequence of this approach is that it is orthogonal to the community competition for scaling proofs to large networks: usually the property Q_1 does not restrict the size of neural networks, but rather points to their structural properties. Thus, implicitly we quantify over neural networks of any size.

To emulate a property *à la* de Maria et al., in [6] we defined a general network monotonicity property: *any fully connected network with positive weights is monotone, in the sense that, given increasing positive inputs, its outputs will also increase.* There has been some interest in monotone networks in the literature [19,21]. Our experiments show that Imandra can prove such properties by induction on the networks' structure almost automatically (with the help of a handful of auxiliary lemmas). And the proofs easily go through for both quantised and real-valued neural networks.[1]

---

[1] Note that in these experiments, the implementation of weight matrices as lists of lists is implicit – we redefine matrix manipulation functions that are less general but more convenient for proofs by induction.

4   Matrices as Functions

We now return to the verification challenge of ACAS Xu, which we failed to conquer with the inductive matrix representation of the last section. This time we ask whether representing matrices as functions and leveraging Imandra's default proof automation can help.

With this in mind, we redefine matrices as functions from indices to values, which gives constant-time (recursion-free) access to matrix elements:

```
type arg =
  | Rows
  | Cols
  | Value of int * int

type 'a t = arg -> 'a
```

Listing 1.2. Implementation of matrices as functions from indices to values

Note the use of the `arg` type, which treats a matrix as a function evaluating "queries" (e.g., "how many rows does this matrix have?" or "what is the value at index (i, j)?"). This formalisation technique is used as Imandra's logic does not allow function values inside of algebraic data types. We thus recover some functionality given by refinement types in [14].

Furthermore, we can map functions over a matrix or a pair of matrices (using `map2`), transpose a matrix, construct a diagonal matrix etc. without any recursion, since we work point-wise on the elements. At the same time, we remove the need for error tracking to ensure matrices are of the correct size: because our matrices are total functions, they are defined everywhere (even outside of their stated dimensions), and we can make the convention that all matrices we build are valid and sparse by construction (with default 0 outside of their dimension bounds).

The resulting function definitions are much more succinct than with lists of lists; take for instance `map2`:

```
let map2 (f: 'a -> 'b -> 'c) (m: 'a t) (m': 'b t) : 'c t =
  function
    | Rows -> rows m
    | Cols -> cols m
    | Value (i,j) -> f (m (Value (i,j))) (m' (Value (i,j)))
```

This allows us to define fully connected layers:

```
let fc (f: 'a -> 'b) (weights: 'a Matrix.t) (input: 'a Matrix
    .t) =
  let open Matrix in
  function
    | Rows            -> 1
```

```
| Cols              -> rows weights
| Value (0, j) ->
  let input' = add_weight_coeff input in
  let weights_row = nth_row weights j in
  f (dot_product weights_row input')
| Value _          -> 0
```

As the biases are included in the `weights` matrix, `add_weight_coeff` prepends a column with coefficients 1 to the input so that they are taken into account.

For full definitions of matrix operations and layers, the reader is referred to [6], but we will give some definitions here, mainly to convey the general style (and simplicity!) of the code. Working with the ACAS Xu networks [12], a script transforms the original networks into sparse functional matrix representation. For example, layer 5 of one of the networks we used is defined as follows:

```
let layer5 = fc relu (
  function
  | Rows -> 50
  | Cols -> 51
  | Value (i,j) -> Map.get (i,j) layer5_map)

let layer5_map =
  Map.add (0,0) (1) @@
  Map.add (0,10) (-1) @@
  Map.add (0,29) (-1) @@
  ...
  Map.const 0
```

The sparsity effect is achieved by *pruning* the network, i.e. removing weights that have the smallest impact on the network's performance. The weight's magnitude is used to select those to be pruned. This method, though rudimentary, is considered a reasonable pruning technique [15]. We do this mainly in order to reduce the amount of computation Imandra needs to perform, and to make the verification problem amenable to Imandra.

With this representation, we are able to verify the properties described in [12] on some of the pruned networks (see Table 1). This is a considerable improvement compared to the previous section, where the implementation did not allow to verify even pruned networks. It is especially impressive that it comes "for free" by simply changing the underlying matrix representations.

Several factors played a role in automating the proof. Firstly, by using maps for the large matrices, we eliminate all recursion (and large case-splits) except for matrix folds (which now come in only via the dot product), which allowed Imandra to expand the recursive matrix computations on demand. Secondly, Imandra's native simplifier contributed to the success. It works on a DAG representation of terms and speculatively expands instances of recursive functions, only as they are (heuristically seen to be) needed. Incremental congruence closure and simplex data structures are shared across DAG nodes, and symbolic execution results are memoised. The underlying `Map.t` components of the functions are reasoned about using a decision procedure for the theory of arrays. Informally speaking, Imandra works lazily expanding out the linear algebra as it is needed, and eagerly sharing information over the DAG. Contrast this approach with that of Reluplex which, informally, starts with the linear algebra fully expanded, and then works to derive laziness and sharing.

Although Imandra's simplifier-based automation above could give us results which `blast` could not deliver for the same network, it still did not scale to the original non-quantised (dense) ACAS Xu network. Contrast this with domain-specific verifiers such as Marabou which are able to scale (modulo potential floating point imprecision) to the full ACAS Xu. We are encouraged that the results of this section were achieved without tuning Imandra's generic proof automation strategies, and hopeful that the development of neural-network specific tactics will help Imandra scale to such networks in the future.

5   Real-Valued Matrices; Records and Arrays

It is time we turn to the question of incorporating real values into matrices. Section 3 defined matrices as lists of lists; and that definition in principle worked for both integer and real-valued matrices. However, we could not use `[@@blast]` to automate proofs when real values were involved; this meant we were restricted to verifying integer-valued networks. On the other hand, the matrix as function implementation extends to proofs with real valued matrices, however it is not a trivial extension. In the functional implementation, the matrix's value must be of the same type as its dimensions (Listing 1.2). Thus, if the matrix elements are real-valued, then in this representation the matrix dimensions will be real-valued as well. This, it turns out, is not trivial to deal with for functions which do recursion along matrix dimensions.

To simplify the code and the proofs, three potential solutions were considered:

- Using an algebraic data type for results of matrix queries: this introduces pattern matching in the implementation of matrix operations, which Sect. 3 taught us to avoid.
- Define a matrix type with real-valued dimensions and values: this poses the problem of proving the function termination when using matrix dimensions in recursion termination conditions.
- Use *records* to provide polymorphism and allow matrices to use integer dimensions and real values.

This section focuses on these three alternatives.

5.1   Algebraic Data Types for Real-Valued Matrices

The first alternative is to introduce an algebraic data type that allows the matrix functions to return either reals or integers.

Table 1. Results of experiments ran on the properties and networks from the ACAS Xu benchmark [12]. The CheckINN verifications were run with 90% of the weights pruned, on virtual machines with four 2.6 GHz Intel Ice Lake virtual processors and 16GB RAM. Timeout was set to 5 h

Property	Result	CheckINN: Pruned Networks		Reluplex: Full ACAS Xu Networks	
		Quantity	Time (s)	Quantity	Time (s)
$\phi 1$	SAT	20	13387	0	
	UNSAT	0		41	394517
	TIMEOUT	24		4	
$\phi 2$	SAT	7	2098	35	82419
	UNSAT	2	896	1	463
	TIMEOUT	26		4	
$\phi 3$	SAT	39	10453	35	82419
	UNSAT	0		1	463
	TIMEOUT	2		4	
$\phi 4$	SAT	36	21533	0	
	UNSAT	0		32	12475
	TIMEOUT	5		0	
$\phi 5$	SAT	1	98	0	
	UNSAT	0		1	19355
$\phi 6$	SAT	1	98	0	
	UNSAT	0		1	180288
$\phi 7$	TIMEOUT	1		1	
$\phi 8$	SAT	0		1	40102
	TIMEOUT	1		0	
$\phi 9$	SAT	1	109	0	
	UNSAT	0		1	99634
$\phi 10$	SAT	0		0	
	UNSAT	1	637	1	19944
	TIMEOUT	0		0	

```
type arg =
  | Rows
  | Cols
  | Value of int * int
  | Default

type 'a res =
  | Int of int
```

```
| Val of 'a

type 'a t = arg -> 'a res
```

This allows a form of polymorphism, but it also introduces pattern matching each time we query a value from the matrix. For instance, in order to use dimensions as indices to access a matrix element we have to implement the following `nth_res` function:

```
let nth_res (m: 'a t) (i: 'b res) (j: 'c res): 'a res = match
    (i, j) with
  | (Int i', Int j') -> m (Value (i', j'))
  | _                -> m Default
```

The simplicity and efficiency of the functional implementation is lost.

5.2   Real-Valued Matrix Indices

We then turn to using real numbers to encode matrix dimensions. The implementation is symmetric to the one using integers (Listing 1.2):

```
type arg =
  | Rows
  | Cols
  | Value of real * real

type 'a t = arg -> 'a
```

A problem arises in recursive functions where matrix dimensions are used as decrementors in stopping conditions, for instance in the `fold_rec` function used in the implementation of the folding operation.

```
let rec fold_rec f cols i j (m: 'a t) =
  let dec i j =
    if j <=. 0. then (i-.1.,cols) else (i,j-.1.)
  in
  if (i <=. 0. && j <=. 0.) || (i <. 0. || j <. 0.) then (
    m (Value (i,j))
  ) else (
    let i',j' = dec i j in
    f (m (Value (i,j))) (fold_rec f cols i' j' m)
  )

let fold (f : 'a -> 'b -> 'b) (m: 'a t) : 'b =
  let rows = m Rows -. 1. in
  let cols = m Cols -. 1. in
  fold_rec f cols rows cols m
```

Imandra only accepts definitions of functions for which it can prove termination. The dimensions being real numbers prevents Imandra from being able to prove termination without providing a custom measure. In order to define this measure, we need to connect the continuous world of reals with the discrete world of integers (and ultimately ordinals) for which we have induction principles. We chose to develop a `floor` function that allows Imandra to prove termination with reals.

To prove termination of our `fold_rec` function recursing along reals, we define an `int_of_real : real -> int` function in Imandra, using a subsidiary `floor : real -> int -> int` which computes an integer floor of a real by "counting up" using its integer argument. In fact, as matrices have non-negative dimensions, it suffices to only consider this conversion for non-negative reals, and we formalise only this. We then have to prove some subsidiary lemmas about the arithmetic of real-to-integer conversion, such as:

```
lemma floor_mono x y b =
  Real.(x <= y && x >= 0. && y >= 0.)
  ==> floor x b <= floor y b
```

```
lemma inc_by_one_bigger_conv x =
  Real.(x >= 0. ==> int_of_real (x + 1.0) > int_of_real x)
```

Armed with these results, we can then prove termination of `fold_rec` and admit it into Imandra's logic via the ordinal pair measure below:

```
[@@measure Ordinal.pair
           (Ordinal.of_int (int_of_real i))
           (Ordinal.of_int (int_of_real j))]
```

Extending the functional matrix implementation to reals was not trivial, but it did have a real payoff. Using this representation, we were able to verify real-valued versions of the pruned ACAS Xu networks! In both cases of integer and real-valued matrices, we pruned the networks to 10% of their original size. So, we still do not scale to the full ACAS Xu challenge. However, the positive news is that the real-valued version of the proofs uses the same waterfall proof tactic of Imandra, and requires no extra effort from the programmer to complete the proof. This result is significant bearing in mind that many functional and higher-order theorem provers are known to have significant drawbacks when switching to real numbers.

From the functional programming point of view, one may claim that this approach is not elegant enough because it does not provide true polymorphism as it encodes matrix dimensions as reals. This motivates us to try the third alternative, using *records* with *maps* to achieve polymorphism.

5.3   Records

Standard OCaml records are available in Imandra, though they do not support functions as fields. This is because all records are data values which must support a computable equality relation, and in general one cannot compute equality on functions. Internally in the logic, records correspond to algebraic data types with a single constructor and the record fields to named constructor arguments. Like product types, records allow us to group together values of different types, but with convenient accessors and update syntax based on field names, rather than position. This offers the possibility of polymorphism for our matrix type.

The approach here is similar to the one in Sect. 4: matrices are stored as mappings between indices and values, which allows for constant-time access to the elements. However, instead of having the mapping be implemented as a function, here we implement it as a Map, i.e. an unordered collection of (key;value) pairs where each key is unique, so that this "payload" can be included as the field of a record.

```
type 'a t = {
  rows: int;
  cols: int;
  vals: ((int*int), 'a) Map.t;
}
```

We can then use a convenient syntax to create a record of this type. For instance, a weights matrix from one of the ACAS Xu networks can be implemented as:

```
let layer6_map =
  Map.add (0,10) (0.05374) @@
  Map.add (0,20) (0.05675) @@
  ...
  Map.const 0.

let layer6_matrix = {
  rows = 5;
  cols = 51;
  vals = layer6_map;
}
```

Note that the matrix dimensions (and the underlying map's keys) are indeed encoded as integers, whereas the weights' values are reals.

Similarly to the previous implementations, we define a number of useful matrix operations which will be used to define general neural network layer functions. For instance, the map2 function is defined thus:

```
let rec map2_rec (m: 'a t) (m': 'b t) (f: 'a -> 'b -> 'c) (
    cols: int) (i: int) (j: int) (res: ((int*int), 'c) Map.t)
    : ((int*int), 'c) Map.t =
    let dec i j =
        if j <= 0 then (i-1, cols) else (i,j-1)
    in
    if i <= 0 && j <= 0 then (
        res
    ) else (
        let (i',j') = dec i j in
        let new_value = f (nth m (i',j')) (nth m' (i', j')) in
        let res' = Map.add' res (i',j') new_value in
        map2_rec m m' f cols i' j' res'
    )
[@@adm i,j]

let map2 (f: 'a -> 'b -> 'c) (m: 'a t) (m': 'b t) : 'c t =
    let rows = max (m.rows) (m'.rows) in
    let cols = max (m.cols) (m'.cols) in
    let vals = map2_rec m m' f cols rows cols (Map.const 0.) in
    {
        rows = rows;
        cols = cols;
        vals = vals;
    }
```

Compared to the list implementation, this implementation has the benefit of providing constant-time access to matrix elements. However, compared to the implementation of matrices as functions, it uses recursion to iterate over matrix values which results in a high number of case-splits. This in turn results in lower scalability. Compared to the previous section's results, none of the verification tests on pruned ACAS Xu benchmarks that terminated with the functional matrix implementation terminated with the records implementation.

Moreover, we can see in the above function definition that we lose considerable conciseness and readability.

In the end, the main interest of this implementation is its offering polymorphism. In all other regards, the functional implementation seems preferable.

6   Conclusions

Functional programming languages that are tightly coupled with automated reasoning capabilities, like Imandra, offer us the possibility to verify and perform inference with neural networks, which the library CheckINN aims to do. To that aim, implementing matrices and matrix operations is important. We have shown different implementations of matrices and how each implementation influences verification in Imandra.

This study has three positive conclusions:

- Imandra's language is sufficiently flexible to give rise to implementations of several choices of matrix in the CheckINN library. Its proof heuristics adapt smoothly to these different implementations, with very little hints needed to figure out the appropriate proof strategy (induction, waterfall or SAT/SMT proving).
- this flexibility bears benefits when it comes to diversifying the range of NN properties we verify: thus, matrices as lists made possible proofs of higher-order properties by induction, whereas matrices as functions were more amenable to automated proofs in SAT/SMT solving style.
- the transition from integer-valued to real-valued NNs is possible in Imandra. This transition itself opens several choices for matrix representations. However, if the matrix representation is optimal for the task at hand, Imandra takes care of completing the proofs with reals and adapts its inductive waterfall method to the new data type automatically. This is a positive lesson to learn, as this is not always given in functional theorem provers.

The main drawback is our failure to scale to the full ACAS Xu problem regardless of the matrix implementation choice in CheckINN [6]. However, it may not come as a great surprise, as general-purpose SMT solvers do not scale to the problem, either. It took domain-specific algorithms such as *ReluPlex* and special-purpose solvers such as *Marabou* to overcome the scaling problem [12,13]. This suggests future solutions that are somewhat orthogonal to the choice of the matrix representation:

- interface with Marabou or other specialised NN solvers in order to scale;
- work on a set of Imandra's native proof heuristics and tactics, tailored specifically to Imandra's NN formalisations.

In addition, evaluating CheckINN against other benchmarks would allow to assess more accurately its scalability on different problems, e.g. robustness verification of image classification networks on the MNIST dataset [1] or range analysis of randomly generated networks [7]. We leave these as future work.

These conclusions provide a strong foundation to further develop the CheckINN library, as its aim is to offer verification of a wide array of neural network properties and we have shown that the choice of matrix implementation eventually influences the range of verifiable properties.

Finally, we believe that the methods we described could be useful in other theorem provers (both first- and higher-order) that combine functional programming and automated proof methods, such as ACL2, PVS, Isabelle/HOL and Coq. For example, in all these systems functions defining matrix operations (e.g., convolution) over lists are often more complex compared to their counterparts over matrices represented as functions, which can benefit from non-recursive definitions. Overall, as these various prominent theorem proving systems work ultimately with functional programs over algebraic datatypes like Imandra, our core observations carry over to them in a natural way.

References

1. VNN (2022). https://sites.google.com/view/vnn2022
2. Lee, R., Jha, S., Mavridou, A., Giannakopoulou, D. (eds.): NFM 2020. LNCS, vol. 12229. Springer, Cham (2020). https://doi.org/10.1007/978-3-030-55754-6
3. Bagnall, A., Stewart, G.: Certifying true error: machine learning in Coq with verified generalization guarantees. AAAI **33**, 2662–2669 (2019)
4. Boyer, R.S., Moore, J.S.: A Computational Logic. ACM Monograph Series. Academic Press, New York (1979)
5. Casadio, M., et al.: Neural network robustness as a verification property: a principled case study. In: Computer Aided Verification (CAV 2022). Lecture Notes in Computer Science, Springer, Cham (2022) https://doi.org/10.1007/978-3-031-13185-1_11
6. Desmartin, R., Passmore, G., Komendantskaya, E., Daggitt, M.L.: CNN library in Imandra. https://github.com/aisec-private/ImandraNN (2022)
7. Dutta, S., Jha, S., Sankaranarayanan, S., Tiwari, A.: Output range analysis for deep feedforward neural networks. In: Dutle, A., Muñoz, C., Narkawicz, A. (eds.) NFM 2018. LNCS, vol. 10811, pp. 121–138. Springer, Cham (2018). https://doi.org/10.1007/978-3-319-77935-5_9
8. Gehr, T., Mirman, M., Drachsler-Cohen, D., Tsankov, P., Chaudhuri, S., Vechev, M.T.: AI2: safety and robustness certification of neural networks with abstract interpretation. In: S&P (2018)
9. Grant, P.W., Sharp, J.A., Webster, M.F., Zhang, X.: Sparse matrix representations in a functional language. J. Funct. Program. 6(1), 143–170 (1996). https://doi.org/10.1017/S095679680000160X, https://www.cambridge.org/core/journals/journal-of-functional-programming/article/sparse-matrix-representations-in-a-functional-language/669431E9C12EDC16F02603D833FAC31B, publisher: Cambridge University Press
10. Heras, J., Poza, M., Dénès, M., Rideau, L.: Incidence simplicial matrices formalized in Coq/SSReflect. In: Davenport, J.H., Farmer, W.M., Urban, J., Rabe, F. (eds.) CICM 2011. LNCS (LNAI), vol. 6824, pp. 30–44. Springer, Heidelberg (2011). https://doi.org/10.1007/978-3-642-22673-1_3
11. Huang, X., Kwiatkowska, M., Wang, S., Wu, M.: Safety verification of deep neural networks. In: Computer Aided Verification - 29th International Conference, CAV 2017, Heidelberg, Germany, July 24–28, 2017, Proceedings, Part I. Lecture Notes in Computer Science, vol. 10426, pp. 3–29 (2017)
12. Katz, G., Barrett, C., Dill, D., Ju-lian, K., Kochenderfer, M.: Reluplex: an Efficient SMT solver for verifying deep neural networks. In: CAV (2017)
13. Katz, G., et al.: The marabou framework for verification and analysis of deep neural networks. In: Dillig, I., Tasiran, S. (eds.) CAV 2019. LNCS, vol. 11561, pp. 443–452. Springer, Cham (2019). https://doi.org/10.1007/978-3-030-25540-4_26
14. Kokke, W., Komendantskaya, E., Kienitz, D., Atkey, R., Aspinall, D.: Neural networks, secure by construction. In: Oliveira, B.C.S. (ed.) APLAS 2020. LNCS, vol. 12470, pp. 67–85. Springer, Cham (2020). https://doi.org/10.1007/978-3-030-64437-6_4
15. LeCun, Y., Denker, J., Solla, S.: Optimal Brain Damage. In: Advances in Neural Information Processing Systems, vol. 2. Morgan-Kaufmann (1989). https://papers.nips.cc/paper/1989/hash/6c9882bbac1c7093bd25041881277658-Abstract.html
16. De Maria, E., et al.: On the use of formal methods to model and verify neuronal archetypes. Front. Comput. Sci. **16**(3), 1–22 (2022). https://doi.org/10.1007/s11704-020-0029-6

17. Passmore, G., et al.: The Imandra automated reasoning system (System Description). In: Peltier, N., Sofronie-Stokkermans, V. (eds.) IJCAR 2020. LNCS (LNAI), vol. 12167, pp. 464–471. Springer, Cham (2020). https://doi.org/10.1007/978-3-030-51054-1_30

18. Passmore, G.O.: Some lessons learned in the industrialization of formal methods for financial algorithms. In: Huisman, M., Păsăreanu, C., Zhan, N. (eds.) FM 2021. LNCS, vol. 13047, pp. 717–721. Springer, Cham (2021). https://doi.org/10.1007/978-3-030-90870-6_39

19. Sill, J.: Monotonic Networks. California Institute of Technology, Pasadena (1998)

20. Singh, G., Gehr, T., Püschel, M., Vechev, M.T.: An abstract domain for certifying neural networks. PACMPL 3(POPL), 41:1–41:30 (2019). https://doi.org/10.1145/3290354

21. Wehenkel, A., Louppe, G.: Unconstrained monotonic neural networks. In: Advances in Neural Information Processing Systems : Annual Conference on Neural Information Processing Systems 2019, IPS 2019, December 8–14, 2019, Vancouver, BC, Canada 32, pp. 1543–1553 (2019)

22. Wood, J.: Vectors and Matrices in Agda (Aug 2019). https://personal.cis.strath.ac.uk/james.wood.100/blog/html/VecMat.html

Self-correcting Neural Networks for Safe Classification

Klas Leino[1], Aymeric Fromherz[2], Ravi Mangal[1(✉)], Matt Fredrikson[1],
Bryan Parno[1], and Corina Păsăreanu[1,3]

[1] Carnegie Mellon University, Pittsburgh, PA 15213, USA
kleino@andrew.cmu.edu rmangal@andrew.cmu.edu mfredrik@cmu.edu
parno@cmu.edu pcorina@andrew.cmu.edu
[2] Inria Paris, 75012 Paris, France
aymeric.fromherz@inria.fr
[3] NASA Ames, Moffett Field, CA 94035, USA

Abstract. Classifiers learnt from data are increasingly being used as components in systems where safety is a critical concern. In this work, we present a formal notion of safety for classifiers via constraints called *safe-ordering constraints*. These constraints relate requirements on the order of the classes output by a classifier to conditions on its input, and are expressive enough to encode various interesting examples of classifier safety specifications from the literature. For classifiers implemented using neural networks, we also present a run-time mechanism for the enforcement of safe-ordering constraints. Our approach is based on a *self-correcting layer*, which provably yields safe outputs regardless of the characteristics of the classifier input. We compose this layer with an existing neural network classifier to construct a *self-correcting network* (SC-Net), and show that in addition to providing safe outputs, the SC-Net is guaranteed to preserve the classification accuracy of the original network whenever possible. Our approach is independent of the size and architecture of the neural network used for classification, depending only on the specified property and the dimension of the network's output; thus it is scalable to large state-of-the-art networks. We show that our approach can be optimized for a GPU, introducing run-time overhead of less than 1 ms on current hardware—even on large, widely-used networks containing hundreds of thousands of neurons and millions of parameters. Code available at github.com/cmu-transparency/self-correcting-networks.

Keywords: Safety · Run-time enforcement · Machine learning · Neural networks · Verification

O. Isac et al. (Eds.): NSV 2022/FoMLAS 2022, LNCS 13466, pp. 96–130, 2022.
https://doi.org/10.1007/978-3-031-21222-2_7

1   Introduction

Classifiers in the form of neural networks are being deployed as components in many safety- and security-critical systems, such as autonomous vehicles, banking systems, and medical diagnostics. A well-studied example is the ACAS Xu networks [20], which provide guidance to an airborne collision avoidance system for commercial aircraft. Unfortunately, standard network training approaches will typically produce models that are accurate but unsafe [29,33]. The ACAS Xu networks, in particular, have been shown [21] to violate safety properties formulated by the developers [20].

What are safety properties for classifiers? Classifiers implemented as neural networks are programs of type $\mathbb{R}^n \to \mathbb{R}^m$, where typically the index of the maximum element of the output m-tuple represents the predicted class. Such classifiers also give an order on the classes, from most likely to least, represented by the order on indices induced by sorting the elements (also referred to as *logits*) of the tuple, and in a variety of domains, systems with classifier components may use this ordering, in addition to the top predicted class, for downstream decision-making.

The ACAS Xu classifiers are an example of a domain where ordering matters. They map sensor readings about the physical state of the aircraft to horizontal maneuver advisories. The sensor readings are imperfect, and the system only has access to a distribution function (or, alternatively, a set of samples) that assigns probability $b(s)$ to being in state s. To issue a maneuver guidance in real time, at each time-step, the system finds the maneuver that maximizes $\sum_s Q(s)_a b(s)$ where $Q(s)_a$ is the value assigned by the neural classifier to maneuver a in state s. As a consequence, the order of the classes, in addition to the top class, are relevant when defining safety properties of ACAS Xu networks.

Another example domain is image classification, where popular datasets, such as CIFAR-100 and ImageNet, have classes with hierarchical structure (e.g., CIFAR-100 has 100 classes with 20 superclasses). Consider a client of an image classifier that averages the logit values over a number of samples for classes that appear in top-k positions and chooses the class with the highest average logit value, due to imperfect sensor information. A reasonable safety property is to require that the chosen class shares its superclass with at least one of the top-1 predictions. This in turn requires reasoning over the order of the classes, and not just the top class.

More generally, the ordering of the logits conveys information about the neural classifier's 'belief' in what the true class is. Under this interpretation, it is natural to express safety constraints on the class order. On the other hand, the exact logit values may be less meaningful, given the approximate nature of neural networks and the fact that logit values are not typically calibrated to any particular value.

Motivated by these observations, we define safety property specifications for classifiers via constraints that we refer to as *safe-ordering constraints*. We argue that these constraints are general enough to encompass the meaningful safety specifications defined for the ACAS Xu networks [21], as well as those used in other safety verification and repair efforts [29,40]. Formally, safe-ordering constraints can specify non-relational safety properties [8] of the form $P \implies Q$, where P is a precondition, expressed as a decidable formula over the classifier's input, and Q is a postcondition, expressed as a statement over its output in the theory of totally ordered sets.

We note that many safety specifications provided by experts are *underconstrained* [21]; i.e., they say what the classifier *should not do* but not what it *should* do. As a specific example, one of the ACAS Xu safety properties roughly states that if an oncoming aircraft is directly ahead and is moving toward our aircraft, then the clear-of-conflict advisory should not have the maximal output from the network[1]. The decision as to what output should have the maximal value must be determined by learning from the input-output examples in the training data. Thus, even when safety specifications are provided, one still needs to perform training based on labeled data to build the classifier, whose performance is measured by computing its accuracy on a separate test set.

Enforcing Safe-Ordering Constraints. Standard approaches for learning neural classifiers will typically produce models that are accurate but unsafe [29,33]. As a result, considerable work has studied the safety of neural networks in general [3, 11,12,15,19,33,34,39], and the ACAS Xu networks in particular [21,29,33,40, 45].

Some approaches use abstract interpretation [15,39] or SMT solving [21] to verify safety properties of networks trained using standard techniques. Unfortunately, the scalability of these techniques remains a serious challenge for most neural-network applications. Furthermore, post-training verification does not address the problem of constructing safe networks to begin with. Retraining the network when verification fails is prohibitively expensive for modern networks [7,41,42], with no guarantees that the train-verify-train loop will terminate. On the other hand, approaches based on statically repairing the network can damage its accuracy (i.e., frequency of the top predicted class matching the 'true' class) on inputs outside the scope of a given safety specification [40]. An alternate approach is to change the learning algorithm such that it provably produces safe networks [29], but such approaches may not converge during training, thus not being able to provide a safety guarantee for the analyzed networks.

In contrast to these previous works, we propose a lightweight, run-time technique for ensuring that neural classifiers are *guaranteed* to satisfy their safe-ordering specifications and at the same time maintain the network's accuracy. Specifically, we describe a program transformer that, given a neural architecture f_θ (parameterized by θ) and a set of safe-ordering constraints Φ, produces a new architecture f_θ^Φ that satisfies the conjunction of Φ for all parameters θ. Viewing the neural network as a composition of layers, our transformer appends a dif-

---

[1] While [20] used the convention that the index of the minimal element of ACAS Xu networks is the top predicted advisory, in this paper we will use the more common convention of the maximal value's index.

ferentiable *self-correcting layer* (SC-Layer) to f_θ. This layer encodes a dynamic *check-and-correct* mechanism, so that when $f_\theta(x)$ violates Φ, the SC-Layer modifies the output to ensure safety. Differentiability of the mechanism also opens the possibility for the training procedure to take self-correction into account during training so that safer and more accurate models can be built, reducing the need for the run-time correction.

Consider again the ACAS Xu networks. Ideally, before deploying the system, we would like to certify that the trained neural classifiers meet their safety specifications. Since the training algorithms are not guaranteed to produce safe classifiers [29,33], and the train-verify-train loop may not terminate, one is likely to be forced to deploy uncertified classifiers. A run-time mechanism that flags safety violations can provide some assurance, but for a real-time, unmanned system like ACAS Xu, throwing exceptions during operation and aborting the computation is not acceptable. Instead, to ensure safe operation without interruptions, we propose to correct the outputs of the classifier whenever necessary.

Our approach is similar in spirit to those that dynamically correct errors in long-running programs caused by traditional software issues like division-by-zero, null dereference, and others [5,22,30,36–38], as well as dynamic check-and-correct mechanisms employed by controllers, referred to as shields [2,6,46]. A check-and-correct mechanism may be impractical for arbitrary classifier safety specifications, as they may require solving arbitrarily complex constraint-satisfaction problems. We show that this is not the case for safe-ordering constraints, and that the solver needed for these constraints can be efficiently embedded in the correction layer.

We note that when correcting the neural network output to enforce safety, we still need to preserve its accuracy. To address the issue, we define a property, *transparency*, which ensures that the correction mechanism has no negative impact on the network's accuracy. Transparency requires that the predicted top class of the original network f_θ be retained whenever it is consistent with at least one ordering allowed by Φ. However, if Φ is inconsistent with the "correct" class specified by the data, then it is impossible for the network to be safe without harming accuracy, and the correction prioritizes safety. We prove that our SC-Layer guarantees transparency. More generally, our correction mechanism tries to retain as much of the original class order as possible.

Finally, while the SC-Layer achieves safety without negatively impacting accuracy, it necessarily adds computational overhead each time the network is executed. We design the SC-Layer, including the embedded constraint solver, to be both vectorized and differentiable, allowing the efficient implementation of our approach within popular neural network frameworks. We also present experiments that evaluate how the overhead is impacted by several key factors. We show that the cost of the SC-Layer depends solely on Φ and the length m of the output vector, and thus, is *independent* of the size or complexity of the underlying neural network. On three widely-used benchmark datasets (ACAS Xu [21], Collision Detection [12], and CIFAR-100 [24]), we show that this overhead is small in real terms (0.26–0.82 ms), and does not pose an impediment to practical adoption. In fact, because the overhead is independent of network size, its impact is less noticeable on larger networks, where the cost of evaluating the original classifier may come to dominate that of the correction. To further characterize the role of Φ and m, we use synthetic data and random safe-ordering constraints to isolate the effects that the postcondition complexity and number of classes have on network run time. While these structural traits of the specified safety constraint can impact run time—the satisfiability of general ordering constraints is NP-complete [16]—our results suggest it will be rare in practice.

Hence, the main contributions of our work are as follows:

- We define *safe-ordering constraints*, as a generic way of writing safety specifications for neural network classifiers.
- We present a method for transforming feed-forward neural network architectures into safe-by-construction versions that are guaranteed to *(i)* satisfy a given set of safe-ordering constraints, and *(ii)* preserve or improve the empirical accuracy of the original model.
- We show that the SC-Layer can be designed to be both fully-vectorized and differentiable, which enables hardware acceleration to reduce run-time overhead, and facilitates its use during training.
- We empirically demonstrate that the overhead introduced by the SC-Layer is small enough for its deployment in practical settings.

2   Problem Setting

In this section, we formalize the concepts of *safe-ordering constraints* and *self-correction*. We begin by presenting background on neural networks and an illustrative application of safe-ordering constraints. We then formally define the problem we aim to solve, and introduce a set of desired properties for our self-correcting transformer.

2.1   Background

Neural Networks. A neural network, $f_\theta : \mathbb{R}^n \to \mathbb{R}^m$, is a total function defined by an *architecture*, or composition of linear and non-linear transformations, and a set of *weights*, θ, parameterizing its linear transformations. As neither the details of a network's architecture nor the particular valuation of its weights are relevant to much of this paper, we will by default omit the subscript θ, and treat f as a black-box function. Neural networks are used as classifiers by extracting *class predictions* from the output $f(x) : \mathbb{R}^m$, also called the *logits* of a network. Given a neural network f, we use the upper-case F to refer to the corresponding neural classifier that returns the top class: $F = \lambda x.\, \mathrm{argmax}_i \{f_i(x)\}$. For our purposes, we will assume that argmax returns a single index, $i^* \in [m]^2$; ties may be broken arbitrarily.

---

[2] $[m] := \{0, \ldots, m-1\}$.

ACAS Xu: An Illustrative Example. We use ACAS Xu [20] as a running example to illustrate key aspects of the problem that our approach solves. The Airborne Collision Avoidance System X (ACAS X) [23] is a family of collision avoidance systems for both manned and unmanned aircraft. ACAS Xu, the variant for unmanned aircraft, is implemented as a large (2GB) numeric lookup table mapping the physical state of the aircraft and a neighboring object (an *intruder*) to horizontal maneuver advisories. The lookup table is indexed on the distance (ρ) between the aircraft and the intruder, the relative angle (θ) from the aircraft to the intruder, the angle (ψ) from the intruder's heading to the aircraft's heading, the speed of the aircraft (v_{own}), and of the intruder (v_{int}), and the time (τ) until loss of vertical separation. The possible advisories are either that no change is needed (or clear-of-conflict, COC), that the aircraft should steer weakly to the left, weakly to the right, strongly to the left, or strongly to the right.

As the table is too large for many unmanned avionics systems, [20] proposed the use of neural networks as a compressed, functional representation of the lookup table. The networks proposed by [20] are functions $f : \mathbb{R}^5 \to \mathbb{R}^5$; the value τ is discretized and 45 different neural networks are constructed, one for each combination of the previous advisory (a_{prev}) and discretized value of τ. Note that while the neural representation of the lookup table is an effective way to encode it on resource-constrained avionics systems, they are necessarily an approximation of the desired functionality, and may thus introduce unsafe behavior [21, 29, 33, 40, 45]. To address this, [21] proposed 10 safety properties, which capture requirements such as, "If the intruder is directly ahead and is moving towards the ownship, the score for *COC* will *not* be maximal." Our goal is to construct networks that are guaranteed to satisfy specifications like these.

2.2   Problem Definition

Definition 1 presents the safe-ordering constraints that we consider throughout the rest of the paper. Intuitively, they correspond to constraints on the relative ordering of a network's output values (a postcondition) with a predicate on the corresponding input (a precondition). As we will see in later sections, the precondition does not need to belong to a particular theory, and need only come with an effective procedure for deciding new instances.

Definition 1 (Safe ordering constraint). *Given a neural network $f : \mathbb{R}^n \to \mathbb{R}^m$, a safe-ordering constraint, $\phi = \langle P, Q \rangle$, is a precondition, P, consisting of a decidable proposition over $\mathbb{R}^n$, and a postcondition, Q, given as a Boolean combination of order relations between the real components of $\mathbb{R}^m$.*

$$
\begin{array}{ll}
precondition & P := decidable\ proposition \\
ordering\ literal & q := y_i < y_j \ (0 \le i, j < m) \\
ordering\ constraint & Q := q \mid Q \wedge Q \mid Q \vee Q \\
safe\text{-}ordering\ constraint & \phi := \langle P, Q \rangle \\
set\ of\ constraints & \Phi := \cdot \mid \phi, \Phi
\end{array}
$$

Assuming a function, $\mathtt{evalP}\!:\!\mathbb{R}^n \to \mathtt{bool}$, *that decides* P *given* $x \in \mathbb{R}^n$, *notated as* $P(x)$, *and a similar* $\mathtt{eval}$ *function for* Q, *we say* f *satisfies safe-ordering constraint* ϕ *at* x *iff* $P(x) \implies Q(f(x))$. *We use the shorthand* $\phi(x, f(x))$ *to denote this; and given a set of constraints* Φ, *we write* $\Phi(x, f(x))$ *to denote* $\forall \phi \in \Phi . \phi(x, f(x))$ *and* $\Phi(x)$ *to denote* $\bigwedge_{\langle P_i, Q_i \rangle \in \Phi \mid P_i(x)} Q_i$.

Two points about our definition of safe-ordering constraints bear mentioning. First, although postconditions are evaluated using the inequality relation from real arithmetic, we assume that $\forall x . i \neq j \implies f_i(x) \neq f_j(x)$, and thus specifically exclude equality comparisons between the output components. This is a realistic assumption in nearly all practical settings, and in cases where it does not hold, can be resolved with arbitrary tie-breaking protocols that perturb $f(x)$ to remove any equalities. Second, we omit explicit negation from our syntax, as it can be achieved by swapping the positions of the affected order relations; i.e., $\neg(y_i < y_j)$ is just $y_j < y_i$, as we exclude the possibility that $y_i = y_j$.

Sections 5.3 and 5.4 provide several concrete examples of safe-ordering constraints. Example 1 revisits the safety specification for ACAS Xu that was discussed in the previous section. Notice that this specification is an instance of the situation where *safety need not imply accuracy*, since it does not specify what category *should* be maximal; that choice must be learned from the training data.

Example 1 (Safety need not imply accuracy). Recall the specification described earlier: "If the intruder is directly ahead and is moving towards the ownship, the score for COC will not be maximal." This is a safe-ordering constraint $\langle P, Q \rangle$, where the precondition P is captured as a linear real arithmetic formula given by [21]:

$$P \equiv 1500 \leq \rho \leq 1800 \ \wedge \ -0.06 \leq \theta \leq 0.06 \ \wedge \ \psi \geq 3.10$$
$$\wedge \ v_{own} \geq 980 \ \wedge \ v_{int} \geq 960$$
$$Q \equiv y_0 < y_1 \ \vee \ y_0 < y_2 \ \vee \ y_0 < y_3 \ \vee \ y_0 < y_4$$

In fact, nine of the ten specifications proposed by [21] are safe-ordering constraints. The single exception has a postcondition that places a constant lower-bound on y_0, i.e., a constraint on the logit value. We do not consider such constraints because the exact logit values are often less meaningful than the class order, given the approximate nature of neural networks and the fact that logit values are not typically calibrated. Moreover, the logit values of the network can be freely scaled without impacting the network's behavior as a classifier.

Given a set of safe-ordering constraints, Φ, our goal is to obtain a neural network that satisfies Φ everywhere. In later sections, we show how to accomplish this by describing the construction of a *self-correcting transformer* (Definition 2) that takes an existing, possibly unsafe network, and produces a related model that satisfies Φ at all points. While in practice, a meaningful, well-defined specification Φ should be satisfiable for all inputs, our generic formulation of safe-ordering constraints in Definition 1 does not enforce this restriction; we can, for instance, let $\Phi := \langle \top, y_0 < y_1 \rangle, \langle \top, y_1 < y_0 \rangle$. To account for this, we lift predicates ϕ to operate on $\mathbb{R}^m \cup \{\bot\}$, where $\phi(x, \bot)$ is considered valid for all x.

Definition 2 (Self-correcting transformer). *A self-correcting transformer,* $SC : \Phi \rightarrow (\mathbb{R}^n \rightarrow \mathbb{R}^m) \rightarrow (\mathbb{R}^n \rightarrow (\mathbb{R}^m \cup \{\perp\}))$, *is a function that, given a set of safe-ordering constraints,* Φ, *and a neural network,* $f : \mathbb{R}^n \rightarrow \mathbb{R}^m$, *produces a network, denoted as* $f^\Phi : \mathbb{R}^n \rightarrow (\mathbb{R}^m \cup \{\perp\})$, *that satisfies the following properties:*

(i) Safety: $\forall x . (\exists y. \Phi(x, y)) \implies \Phi(x, f^\Phi(x))$
(ii) Forewarning: $\forall x . (f^\Phi(x) = \perp \iff \forall y . \neg\Phi(x, y))$

In other words, $f^\Phi = SC(\Phi)(f)$ *is safe with respect to* Φ *and produces a non-* $\perp$ *output wherever* $\Phi(x)$ *is satisfiable. We refer to the output of* SC, f^Φ, *as a self-correcting network (SC-Net).*

Definition 2(i) captures the essence of the problem that we aim to solve, requiring that the self-correcting network make changes to its output according to Φ. While allowing it to abstain from prediction by outputting $\perp$ may appear to relax the underlying problem, note that this is only allowed in cases where Φ cannot be satisfied on x: Definition 2(ii) is an equivalence that precludes trivial solutions such as $f^\Phi := \lambda x.\perp$. However, it still allows abstention in exactly the cases where it is needed for principled reasons. A set of safe-ordering constraints may be mutually satisfiable almost everywhere, except in some places; for example: $\Phi := \langle x \leq 0.5, y_0 < y_1 \rangle, \langle x \geq 0.5, y_1 < y_0 \rangle$. In this case, f^Φ can abstain at $x = 0.5$, and everywhere else must produce outputs in $\mathbb{R}^m$ obeying Φ.

While the properties required by Definition 2 are sufficient to ensure a non-trivial, safe-by-construction neural network, in practice, we aim to apply $SC(\Phi)$, which we will write as SC^Φ, to models that *already* perform well on observed test cases, but that still require a safety guarantee. Thus, we wish to correct network outputs without interfering with the existing network behavior when possible, a property we call *transparency* (Property 1).

Property 1 (Transparency). *Let* $SC : \Phi \rightarrow (\mathbb{R}^n \rightarrow \mathbb{R}^m) \rightarrow (\mathbb{R}^n \rightarrow (\mathbb{R}^m \cup \{\perp\}))$ *be a self-correcting transformer. We say that* SC *satisfies transparency if*

$$\forall\Phi . \forall f : \mathbb{R}^n \rightarrow \mathbb{R}^m . \forall x \in \mathbb{R}^n .$$
$$\left(\exists y. \Phi(x, y) \land \underset{i}{\mathrm{argmax}}\{y_i\} = F(x)\right) \implies F^\Phi(x) = F(x)$$

where $F^\Phi(x) := \perp$ *if* $f^\Phi(x) = \perp$ *else* $\mathrm{argmax}_i\{f_i^\Phi(x)\}$. *In other words, SC always produces an SC-Net,* f^Φ, *for which the top class derived from the safe output vectors of* f^Φ *agrees with the top class of the original model whenever possible.*

Property 1 leads to a useful result, namely that whenever Φ is consistent with *accurate* predictions, then the classifier obtained from $SC^\Phi(f)$ is at least as accurate as F (Theorem 2). Formally, we characterize accuracy in terms of agreement with an oracle classifier F^O that "knows" the correct class for each input, so that F is accurate on x if and only if $F(x) = F^O(x)$. We note that accuracy is often defined with respect to a *distribution* of labeled points rather than an oracle; however our formulation captures the key fact that Theorem 2 holds regardless of how the data are distributed.

Theorem 2 (Accuracy Preservation). *Given a neural network, $f : \mathbb{R}^n \to \mathbb{R}^m$, and set of constraints, Φ, let $f^\Phi := SC^\Phi(f)$ and let $F^O : \mathbb{R}^n \to [m]$ be the oracle classifier. Assume that SC satisfies transparency. Further, assume that accuracy is consistent with safety, i.e.,*

$$\forall x \in \mathbb{R}^n . \exists y . \Phi(x,y) \land \operatorname*{argmax}_i \{y_i\} = F^O(x).$$

Then,

$$\forall x \in \mathbb{R}^n . F(x) = F^O(x) \implies F^\Phi(x) = F^O(x)$$

One subtle point to note is that even when Φ is consistent with accurate predictions, it is possible for a network to be accurate yet unsafe at an input. Example 2 describes such a situation. Our formulation of Property 1 is carefully designed to ensure accuracy preservation even in such scenarios.

Example 2 (Accuracy need not imply safety). Consider the property ϕ_2 proposed for ACAS Xu by [21] which says: "If the intruder is distant and is significantly slower than the ownship, the score of the COC advisory should never be minimal." This safe-ordering constraint is applicable for all networks that correspond to $a_{prev} \neq COC$ and is concretely written as follows:

$$P \equiv \rho \geq 55947.691 \land v_{own} \geq 1145 \land v_{int} \leq 60$$
$$Q \equiv y_1 < y_0 \lor y_2 < y_0 \lor y_3 < y_0 \lor y_4 < y_0$$

For some x such that $P(x)$ is true, let us assume that $F^O(x) = 1$ and for a network f, $f(x) = [100, 900, 300, 140, 500]$, so that $F(x) = 1$. Then, f is accurate at x, but the COC advisory receives the minimal score, meaning f is unsafe at x with respect to ϕ_2. If the transformer SR satisfies Property 1, then by Theorem 2, f^{ϕ_2} is guaranteed to be accurate as well as safe at x, since ϕ_2 is consistent with accuracy here (as ϕ_2 does not preclude class 1 from being maximal).

3   Self-correcting Transformer

We describe our self-correcting transformer, SC. We begin with a high-level overview of the approach (Sect. 3.1), and provide algorithmic details in Sect. 3.2. We then provide proofs (Sect. 3.3) and complexity analysis (Sect. 3.4).

3.1   Overview

Our self-correcting transformer, SC, leverages the fact that whenever a safe-ordering constraint is satisfiable at a point, it is possible to bring the network into compliance. Neural networks are typically constructed by composing a sequence of layers; we thus compose an additional *self-correction layer* that operates on the original network's output, and produces a result that will serve as the transformed network's new output. This is reflected in the SC routine in Algorithm 3.1. The original network, f, executes normally, and the self-correction layer subsequently takes both the input x (to facilitate checking the preconditions of Φ) and $y := f(x)$, from which it either abstains (outputs $\perp$) or produces an output that is guaranteed to satisfy Φ.

The high-level workflow of the self-correction layer, SC-Layer, proceeds as follows. The layer starts by checking the input x against each of the preconditions, and derives an *active postcondition*. This is then passed to a solver, which attempts to find the set of orderings that are consistent with the active postcondition. If no such ordering exists, i.e., if the active postcondition is unsatisfiable, then the layer abstains with $\perp$. Otherwise, the layer minimally permutes the indices of the original output vector in order to satisfy the active postcondition while ensuring transparency (Property 1).

3.2   Algorithmic Details of SC-Layer

The core logic of our approach is handled by a self-correction layer, or SC-Layer, that is appended to the original model, and dynamically ensures its outputs satisfy the requisite safety specifications. The procedure followed by this layer, SC-Layer (shown in Algorithm 3.1), first checks if the input x and output y of the base network already satisfy Φ (line 5). If they do, no correction is necessary and the repaired network f^Φ can safely return y. Otherwise, SC-Layer attempts to find a satisfiable ordering constraint that entails the relevant postconditions in Φ (line 8). FindSatConstraint either returns such a term q that consists of a conjunction of ordering literals $y_i < y_j$, or returns $\perp$ whenever no such q exists. When FindSatConstraint returns $\perp$, then SC-Layer does as well (lines 9–10). Otherwise, the constraint identified by FindSatConstraint is used to correct the network's output (line 12), where Correct permutes the logit values in y to arrive at a vector that satisfies q. Note that because q is satisfiable, it is always possible to find a satisfying solution by simply permuting y, because the specific real values are irrelevant, and only their order matters (see Sect. 3.3).

Finding   Satisfiable   Constraints.   Algorithm 3.2   illustrates   the FindSatConstraint procedure. Recall that the goal is to identify a conjunction of ordering literals q that implies the *relevant* postconditions in Φ at the given input x. More precisely, this means that for each precondition P_i satisfied by x, the corresponding postcondition Q_i is implied by q. This is sufficient to ensure that any model y' of q will satisfy Φ at x; i.e., $q(y') \implies \Phi(x, y')$.

To accomplish this, FindSatConstraint first evaluates each precondition, and obtains (line 9) a disjunctive normal form (DNF), Q_x, of the *active postcondition*, defined by $\text{Filter}(\Phi, x) := \bigwedge_{\langle P_i, Q_i \rangle \in \Phi \mid P_i(x)} Q_i$. In practice, we implement a lazy version of ToDNF that generates disjuncts as needed (see Sect. 4),

---

Algorithm 3.1: Self-correcting transformer

Inputs: A set of safety properties, Φ and a network, $f : \mathbb{R}^n \to \mathbb{R}^m$
Output: A network, $f^\Phi : \mathbb{R}^n \to \mathbb{R}^m \cup \{\bot\}$

```
1  SC(Φ , f):
2  │  f^Φ := λ x.SC-Layer(Φ, x, f(x))
3  │  return f^Φ

4  SC-Layer(Φ , x , y):
5  │  if Φ(x, y) then
6  │  │  return y
7  │  else
8  │  │  q := FindSatConstraint(Φ, x, y)
9  │  │  if q = ⊥ then
10 │  │  │  return ⊥
11 │  │  else
12 │  │  │  y' := Correct(q, y)
13 │  │  │  return y'
```

---

as this step may be a bottleneck, and we only need to process each clause individually. At this point, FindSatConstraint could proceed directly, checking the satisfiability of each disjunct in Q_x, and returning the first satisfiable one it encounters. This would be correct, but as we wish to satisfy transparency (Property 1), we first construct an ordered list of the terms in Q_x which prioritizes constraints that maintain the maximal position of the original prediction, $\text{argmax}(y)$ (Prioritize, line 10). Property 3 formalizes the behavior required of Prioritize.

Property 3 (Prioritize). *Given $y \in \mathbb{R}^m$ and a list of conjunctive ordering constraints $\overline{Q}$, the result of* Prioritize($\overline{Q}, y$) *is a reordered list $\overline{Q}' = [\dots, q_i, \dots]$ such that:*

$$\forall\, 0 \leq i, j < |\overline{Q}| \,.\, \underset{i}{\text{argmax}}\{y_i\} \in Roots(\texttt{OrderGraph}(q_i))$$

$$\wedge\, \underset{i}{\text{argmax}}\{y_i\} \notin Roots(\texttt{OrderGraph}(q_j)) \implies i < j$$

where Roots(G) denotes the root nodes of the directed graph G.

The IsSat procedure (invoked on line 12, also shown in Algorithm 3.2) checks the satisfiability of a conjunctive ordering constraint. It is based on an encoding of q as a directed graph, embodied in OrderGraph (lines 1–4), where each component index of y corresponds to a node, and there is a directed edge from i to j if the literal $y_j < y_i$ appears in q. A constraint q is satisfiable if and only if OrderGraph(q) contains no cycles (lines 5–7) [35]. Informally, acyclicity is necessary and sufficient for satisfiability because the directed edges encode

---

Algorithm 3.2: Finding a satisfiable ordering constraint from safe-ordering constraints Φ

Inputs: A set of safe-ordering constraints, Φ, a vector $x : \mathbb{R}^n$, and a vector $y : \mathbb{R}^m$

Output: Satisfiable ordering constraint, q

1 OrderGraph(q):
2 | $V := [m]$
3 | $E := \{(i,j) : y_j < y_i \in q\}$
4 | **return** (V, E)

5 IsSat(q):
6 | $g := $ OrderGraph(q)
7 | **return** $\neg$ContainsCycle(V, E)

8 FindSatConstraint(Φ , x , y):
9 | $Q_x := $ ToDNF(Filter(Φ, x))
10 | $Q_p := $ Prioritize(Q_x, y)
11 | **foreach** $q_i \in Q_p$ **do**
12 | **if** IsSat(q_i) **then**
13 | **return** q_i

14 | **return** $\bot$

Algorithm 3.3: Correction procedure for safe-ordering constraints

---

Inputs: Satisfiable ordering constraint q, a vector $y : \mathbb{R}^m$

Output: A vector $y' : \mathbb{R}^m$

1 Correct(q , y):
2 | $\pi := $ TopologicalSort(OrderGraph(q), y)
3 | $y^s := $ SortDescending(y)
4 | $\forall j \in [m]$. $y'_j := y^s_{\pi(j)}$
5 | **return** y'

---

immediate ordering requirements, and by transitivity, a cycle involving i entails that $y_i < y_i$.

Correcting Violations. Algorithm 3.3 describes the Correct procedure, used to ensure the outputs of the SC-Layer satisfy safety. The inputs to Correct are a satisfiable ordering constraint q, and the output of the original network $y := f(x)$. The goal is to permute y such that the result y' satisfies q, without violating transparency. Our approach is based on OrderGraph, the same directed-graph encoding used by IsSat. It uses a stable topological sort of the graph encoding of q to construct a total order over the indices of y that is consistent with the partial ordering implied by q (line 2). TopologicalSort returns a permutation π, a function that maps indices in y to their rank (or position) in the total order. Formally, TopologicalSort takes as argument a graph $G = (V, E)$, and returns π such that Eq. 1 holds.

$$\forall i, j \in V . (i, j) \in E \implies \pi(i) < \pi(j) \tag{1}$$

Informally, if the edge (i, j) is in the graph, then i occurs before j in the ordering. In general, many total orderings may be consistent, but in order to guarantee transparency, TopologicalSort also needs to ensure the following invariant (Property 4), capturing that the maximal index is listed first in the total order if possible.

Property 4. *Given a graph, $G = (V, E)$, and $y \in \mathbb{R}^m$, the result π of* TopologicalSort(G, y) *satisfies*

$$\operatorname*{argmax}_i \{y_i\} \in Roots(G) \implies \pi\left(\operatorname*{argmax}_i \{y_i\}\right) = 0$$

where $Roots(G)$ denotes the root nodes of the directed graph G.

In other words, the topological sort preserves the network's original prediction when doing so is consistent with q. Then, by sorting y in descending order, the sorted vector y^s can be used to construct the final output of Correct, y'. For any index i, we simply set y'_i to the $\pi(i)^{th}$ component of y^s, since $\pi(i)$ gives the desired rank of the ith logit value and components in y^s are sorted according to the component values (line 4). Example 3 shows an example of the complete Correct procedure.

Example 3 (Self-correct). We refer again to the safety properties introduced for ACAS Xu [21]. The postcondition of property ϕ_2 states that the logit score for class 0 (COC) is not minimal, which can be written as the following ordering constraint:

$$Q \equiv y_1 < y_0 \lor y_2 < y_0 \lor y_3 < y_0 \lor y_4 < y_0$$

Suppose that for some input $x \in \mathbb{R}^n$, the active postcondition is equivalent to Q, and that $y = [100, 900, 300, 140, 500]$. Further, suppose that FindSatConstraint has returned $q := y_2 < y_0$, corresponding to the second disjunct of Q (satisfying $q \implies Q$). We then take the following steps according to Correct(q, y):

- First we let $\pi := $ TopologicalSort(OrderGraph$(q), y)$. We note that all vertices of the graph representation of q are roots except for $j = 2$, which has $j = 0$ as its parent. We observe that $\operatorname{argmax}_i\{y_i\} = 1$, which corresponds to a root node; thus by Property 4, $\pi(1) = 0$. Moreover, by our ordering constraint, we also have that $\pi(0) < \pi(2)$. Thus, the ordering π where $\pi(0) = 2$, $\pi(1) = 0$, $\pi(2) = 3$, $\pi(3) = 4$, and $\pi(4) = 1$ is a possible result of TopologicalSort, which we will assume for this example.
- Next we obtain by a descending sort that $y^s = [900, 500, 300, 140, 100]$.
- Finally we obtain y' by indexing y^s by the inverse of π, i.e., $y'_j = y^s_{\pi(j)}$. This gives us $y'_0 = y^s_2 = 300$, $y'_1 = y^s_0 = 900$, $y'_2 = y^s_3 = 140$, $y'_3 = y^s_4 = 100$, and $y'_4 = y^s_1 = 500$, resulting in a final output of $y' = [300, 900, 140, 100, 500]$, which (i) satisfies Q, and (ii) preserves the prediction of class 1.

3.3 Key Properties

We now provide a brief argument that our SC procedure satisfies two key properties; namely (1) SC is a self-correcting transformer (Definition 2)—i.e., it guarantees that the corrected output will always satisfy the requisite safety properties, unless they are *unsatisfiable*, in which case it returns $\perp$—and (2) SC is transparent (Property 1)—i.e., it does not modify the predicted class (the class with the maximal logit value) unless it is absolutely necessary for safety. Full proofs appear in Appendix A.

Theorem 5 (SC is a self-correcting transformer). SC *(Algorithm 3.1) satisfies conditions* (i) *and* (ii) *of Definition 2.*

This follows from the construction of SC, and relies on a few key properties of FindSatConstraint and Correct. First, whenever FindSatConstraint returns $\perp$, the set of safety constraints, Φ, is *unsatisfiable* on the given input. Second, whenever FindSatConstraint returns some $q \neq \perp$, then q is satisfiable on the given input. Finally, when q is satisfiable, Correct always modifies the output such that it satisfies q. Together, these imply Theorem 5.

In addition to ensuring safe-ordering, SC is transparent (Theorem 6), which recall is a precondition for the accuracy preservation property stated in Theorem 2.

Theorem 6 (Transparency of SC). SC, *the self-correcting transformer described in Algorithm 3.1, satisfies Property 1.*

Clearly, on points where the model naturally satisfies the safety properties, no changes to the output are made and SC is transparent. Otherwise, we rely on a few key details of our construction to achieve transparency.

We begin with the observation that whenever the network's predicted top class is a root of the graph encoding of a satisfiable postcondition, q, there exists an output that satisfies q while preserving the predicted top class. Intuitively, this follows because the partial ordering admits *any* of the root nodes to appear first in the total ordering.

With this in mind, we recall that FindSatConstraint searches potential solutions according to Prioritize, which prefers all solutions in which the predicted top class appears as a root node over any in which it does not. Thus, Prioritize will return a solution that is consistent with preserving the network's original predicted top class whenever possible.

Finally, we design our topological sort to be "stable," such that, among other things, the network's original top prediction will appear first in the total ordering whenever it appears as a root node. More details on our topological sort algorithm and the properties it possesses are given in Sect. B.1.

3.4 Complexity

Given a neural network $f : \mathbb{R}^n \to \mathbb{R}^m$, we define the input size as n and output size as m. Also, assuming that the postconditions Q_i for all $\langle P_i, Q_i \rangle \in \Phi$

are expressed in DNF, we define the size p_i of a constraint as the number of disjuncts in Q_i and define $\alpha := |\Phi|$, i.e., the number of properties in Φ. Then, the worst-case computational complexity of SC-Layer is given by Eq. 2, where $O(log(m))$ is the complexity of ContainsCycle, $O(mlog(m))$ is the complexity of TopologicalSort, and $\prod_{i=1}^{\alpha} p_i$ is the maximum number of disjuncts possible in Q_x if the postconditions Q_i are in DNF.

$$O\left(log(m) \prod_{i=1}^{\alpha} p_i + m\, log(m)\right) \tag{2}$$

The complexity given by Eq. 2 is with respect to a cost model that treats matrix operations—e.g., matrix multiplication, associative row/column reductions—as constant-time primitives. Crucially, note that the complexity does not depend on the size of the neural network f.

3.5 Differentiability of SC-Layer

One interesting facet of our approach that remains largely unexplored is the differentiability of the SC-Layer. In principle, this opens the door to benefits that could be obtained by training against the corrections made by the SC-Layer. Though we found that a "vanilla" attempt to train with the SC-Layer did not provide clear advantages over appending it to a model post-learning, we believe this remains an interesting future direction to explore. It is conceivable that a careful approach to training SC-Nets could lead to safer and more accurate models, reducing the need for the run-time correction, as the network could learn to use the modifications made by the SC-Layer to its advantage. Furthermore, aspects of the algorithm, including, e.g., the heuristic used to prioritize the search for a satisfiable graph (see **Finding Satisfiable Constraints** in Sect. 3.2), could be parameterized and learned, potentially leading to both accuracy and performance benefits.

4 Vectorizing Self-correction

Widely-used machine learning libraries, such as TensorFlow [1], simplify the implementation of parallelized, hardware-accelerated code by providing a collection of operations on multi-dimensional arrays of uniform type, called *tensors*. One can view such libraries as domain-specific languages that operate primarily over tensors, providing embarrassingly parallel operations like matrix multiplication and associative reduction, as well as non-parallel operations like iterative loops and sorting routines. We use matrix-based algorithms implementing the core procedures used by SC-Layer described in Sect. 3. As we will later see in Sect. 5, taking advantage of these frameworks allows our implementation to introduce minimal overhead, typically fractions of milliseconds. Additionally, it means that SC-Layer can be automatically differentiated, making it fully compatible with training and fine-tuning. One may use an SMT solver like Z3 to implement the procedures used by SC-Layer but we found that making calls to an SMT solver significantly restricts the efficient use of GPUs. Moreover, it is a useful heuristic for the corrected logits to prioritize the original ordering relationships. Encoding this heuristic would require an optimization variant of SMT (Max-SMT). In contrast, our matrix-based algorithms efficiently calculate the corrected output while prioritizing the original class order. We present the algorithmic details in Appendix B.

5   Evaluation

We have shown that self-correcting networks (SC-Nets) provide safety to an existing network without affecting accuracy, as long as safety and accuracy are mutually consistent. This comes with no additional training cost, suggesting that the only potential downside of SC-Nets is the run-time overhead introduced by the SC-Layer. In this section, we present an empirical evaluation of our approach to demonstrate its scalability, and find that the run-time performance is not an issue in practice—overheads range from 0.2–0.8 ms, and scale favorably with the size and complexity of constraints.

We explore the capability of our approach on a variety of domains, demonstrating its ability to solve previously studied safety-verification problems (Sects. 5.1 and 5.2), and its ability to efficiently scale both (i) to large convolutional networks (Sect. 5.3) and (ii) to arbitrary, complex safe-ordering constraints containing disjunctions and overlapping preconditions (Sect. 5.4).

We implemented our approach in Python, using TensorFlow to vectorize our SC-Layer (Sect. 4). All experiments were run on an NVIDIA TITAN RTX GPU with 24 GB of RAM, and a 4.2 GHz Intel Core i7-7700K with 32 GB of RAM.

5.1   ACAS Xu

ACAS Xu [23] is a collision avoidance system that has been frequently studied in the context of neural classifier safety verification [20, 21, 29, 39]. Typically considered for this problem is a family of 45 networks proposed by [20]. [21] proposed 10 safety specifications for this family of networks, which have become standard for research on this problem. We consider 9 of these specifications, which can be expressed as *safe-ordering constraints* (Sect. 2.2).

Each of the 45 networks consists of six hidden dense layers of 50 neurons each. Each network needs to satisfy some subset of the 10 safety constraints; that is, more than one safety constraints may apply to each model, but not all safety constraints apply to each model. A network is considered safe if it satisfies *all* of the relevant safety constraints. Among the 45 networks, [21] reported that 9 networks were already safe after standard training, while 36 were unsafe, exhibiting safety constraint violations.

The data used to train the 45 networks is not publicly available; however, [29] provide a synthetic test set for each network, consisting of 5,000 points uniformly sampled from the specified state space and labeled using the respective network

Table 1. Safety certification results on the **(a)** ACAS Xu [20] and **(b)** Collision Detection [12] datasets. We compare the success rate and accuracy to that of ART [29], a recent safe-by-construction approach. The original network is provided as a baseline. Best results are shown in bold. **(c)** Absolute overhead introduced by the SC-Layer per input.

		Acas Xu	
	method	*safe networks*	*mean accuracy(%)*
36 unsafe nets	original	0 / 36	100.0
	ART	36 / 36	94.4
	SC-Net	**36 / 36**	**100.0**
9 safe nets	original	**9 / 9**	100.0
	ART	9 / 9	94.3
	SC-Net	**9 / 9**	**100.0**

(a)

Collision Detection		
method	*constraints certified*	*accuracy (%)*
original	328 / 500	99.9
ART	481 / 500	96.8
SC-Net	**500 / 500**	**99.9**

(b)

dataset	*overhead (ms)*
ACAS Xu	0.26
Collision Detection	0.58
CIFAR-100 (small CNN)	0.77
CIFAR-100 (ResNet-50)	0.82
Synthetic	0.27

(c)

as an oracle. We note that because this test set is labeled using the original models, the accuracy of each original model on this test set is necessarily 100%.

Table 1a presents the results of applying our SC transformer to each of the 45 provided networks. In particular, we consider the number of networks for which safety can be guaranteed, and the accuracy of the resulting SC-Net. We compare our results to those using ART [29], a recent approach to safe-by-construction learning. ART aims to learn neural networks that satisfy safety specifications expressed using linear real arithmetic constraints. It updates the loss function to be minimized during learning by adding a term, referred to as the *correctness loss*, that measures the degree to which a neural network satisfies or violates the safety specification. A value of zero for the correctness loss ensures that the network is safe. However, there is no guarantee that learning will converge to zero correctness loss, and the resulting model may not be as accurate as one trained with conventional methods.

Because the safety constraints for each network are satisfiable on all points, Definition 2 tells us that safety is guaranteed for all 45 SC-Nets. In this case, we see that ART also manages to produce 45 safe networks after training; however we see that it comes at a cost of nearly 6% points in accuracy, *even on the networks that were already safe*. Meanwhile, transparency (Property 1) tells us that SC-Nets will only see a decrease in accuracy relative to the original network when accuracy is in direct conflict with safety. On the 9 original networks that were reported as safe, clearly no such conflict exists, and accordingly, we see that the corresponding SC-Nets achieve the same accuracy as the original networks (100%). On the 36 unsafe networks, we find again that the SC-Nets achieved 100% accuracy. In this case, it would have been possible that the SC-Nets would have achieved lower accuracy than the original networks, as some of the safety properties have the potential to conflict with accuracy. For example, the post-condition of the property ϕ_8 requires that the predicted maneuver advisory is either to continue straight (COC) or to turn weakly to the left. Thus, correcting ϕ_8 on inputs for which it is violated would necessarily change the network's prediction on those inputs; and, since the labels are derived from the original networks' predictions, this would lead to a drop in accuracy. However, we find that none of the test points include violations of such constraints (even though such violations exist in the space generally [21]), as evidenced by the fact that the SC-Net accuracy remained unchanged.

Table 1c shows the average overhead introduced by applying SC to each of the ACAS Xu networks. We see that the absolute overhead is only ~0.25 ms per instance on average, accounting for less than an 8× increase in prediction time. Note, however, that while our implementation of SC-Layer is optimized for and evaluated using a GPU, the Acas Xu system is expected to be deployed on resource-constrained hardware without access to GPUs [20] which is likely to result in larger overheads.

5.2   Collision Detection

The Collision Detection dataset [12] provides another instance of a safety verification task that has been studied in the prior literature. In this setting, a neural network controller is trained to predict whether two vehicles following curved paths at different speeds will collide. As this is a binary decision task, the network contains two outputs, corresponding to the case of a collision and the case of no collision. [12] proposes 500 safety properties for this task, corresponding to ℓ_∞ *robustness regions* around 500 particular inputs; i.e., property ϕ_i for $i \in \{1, \ldots, 500\}$ corresponds to a point, x_i, and radius, ϵ_i, and is defined according to Eq. 3.

$$\phi_i(x, y) := ||x_i - x||_\infty \leq \epsilon_i \implies y = F(x_i) \tag{3}$$

Such specifications of *local robustness* at fixed inputs can be represented as safe-ordering constraints, where the postcondition of ϕ_i is defined to be $y_0 > y_1$ if $F(x_i) = 0$ and $y_0 < y_1$ if $F(x_i) = 1$.

Table 1b presents the results of applying our SC transformer to the original network provided by [12]. Similarly to before, we consider the number of constraints with respect to which safety can be guaranteed, and the accuracy of the resulting SC-Net, comparing our results to those of ART.

We see in this case that ART was unable to guarantee safety for all 500 specifications. Meanwhile, it resulted in a drop in accuracy of approximately 3% points. On the other hand, it is simple to check that the conjunction of all 500 safety constraints is satisfiable for all inputs; thus, Definition 2 tells us that safety is guaranteed with respect to all properties. Meanwhile SC-Nets impose no penalty on accuracy, as none of the test points violate the constraints.

Table 1c shows the overhead introduced by applying SC to the collision detection model. In absolute terms, we see the overhead is approximately half a millisecond per instance, accounting for under a 3× increase in prediction time.

5.3   Scaling to Larger Domains

One major challenge for many approaches that attempt to verify network safety—particularly post-learning methods—is scalability to very large neural networks. Such networks pose a problem for several reasons. Many approaches analyze the parameters or intermediate neuron activations using algorithms that do not scale polynomially with the network size. This is a practical problem, as large networks in use today contain hundreds of millions of parameters. Furthermore, abstractions of the behavior of large networks may see compounding imprecision in large, deep networks.

Our approach, on the other hand, treats the network as a black-box and is therefore not sensitive to its specifics. In this section we demonstrate that this is borne out in practice; namely the absolute overhead introduced by our SC-Layer remains relatively stable even on very large networks.

For this, we consider a novel set of safety specifications for the CIFAR-100 image dataset [24], a standard benchmark for object recognition tasks. The CIFAR-100 dataset is comprised of 60,000 32 × 32, RGB images categorized into 100 different classes of objects, which are grouped into 20 superclasses of 5 classes each. We propose a set of safe-ordering constraints that are reminiscent of a variant of *top-k accuracy* restricted to members of the same superclass, which has been studied recently in the context of certifying relational safety properties of neural networks [25]. More specifically, we require that if the network's prediction belongs to superclass C_k then the top 5 logit outputs of the network must all belong to C_k. Formally, there are 20 constraints, one for each superclass, where the constraint, ϕ_k for superclass C_k, for $k \in \{1, \ldots, 20\}$, is defined according to Eq. 4. Notice that with respect to these constraints, a standard trained network can be accurate, yet unsafe, even without accuracy and safety being mutually inconsistent.

$$\phi_k(x, y) := F(x) \in C_k \implies \bigwedge_{i,j \,\mid\, i \in C_k, \, j \notin C_k} y_j < y_i \qquad (4)$$

As an example application requiring this specification, consider a client of the classifier that averages the logit values over a number of samples for classes appearing in top-5 positions and chooses the class with the highest average logit value (due to imperfect sensor information - a scenario similar to the ACAS Xu example). A reasonable specification is to require that the chosen class shares its superclass with at least one of the top-1 predictions. We can ensure that this specification is satisfied by enforcing Eq. 4.

Table 1c shows the overhead introduced by applying SC, with respect to these properties, to two different networks trained on CIFAR-100. The first is a convolutional neural network (CNN) that is much smaller than is typically used for vision tasks, containing approximately 1 million parameters. The second is a standard residual network architecture, ResNet-50 [18], with approximately 24 million parameters.

In absolute terms, we see that both networks incur less than 1 ms of overhead per instance relative to the original model (0.77 ms and 0.82 ms, respectively), making SC a practical option for providing safety in both networks. Moreover, the absolute overhead varies only by about 5% between the two networks, suggesting that the overhead is not sensitive to the size of the network. This overhead accounts for approximately a 12× increase in prediction time on the CNN. Meanwhile, the overhead on the ResNet-50 accounts for only a 6× increase in prediction time relative to the original model. The ResNet-50 is a much larger and deeper network; thus its baseline prediction time is longer, so the overhead introduced by the SC-Layer accounts for a smaller fraction of the total computation time. In this sense, our SC transformer becomes relatively *less* expensive on larger networks.

Interestingly, we found that the original network violated the safety constraints on approximately *98% of its inputs*, suggesting that obtaining a violation-free network without SC might prove particularly challenging. Meanwhile, the SC-Net eliminated *all* violations, with *no cost to accuracy*, and less than 1 ms in overhead per instance.

5.4   Handling Arbitrary, Complex Constraints

Safe ordering constraints are capable of expressing a wide range of compelling safety specifications. Moreover, our SC transformer is a powerful, general tool for ensuring safety with respect to arbitrarily complex safe-ordering constraints, comprised of many conjunctive and disjunctive clauses. Notwithstanding, the properties presented in our evaluation thus far have been relatively simple. In this section we explore more complex safe-ordering constraints, and describe experiments that lend insight as to which factors most impact the scalability of our approach.

To this end, we designed a family of synthetic datasets with associated safety constraints that are randomly generated according to several specified parameters, allowing us to assess how aspects such as the number of properties (α), the number of disjunctions per property (β), and the dimension of the output vector (m) impact the run-time overhead. In our experiments, we fix the input

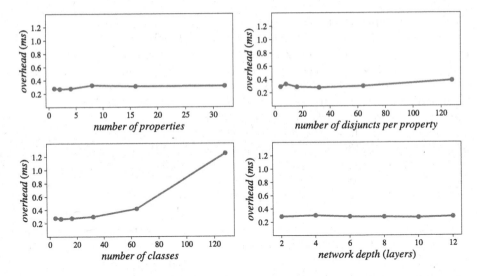

Fig. 1. Absolute overhead in milliseconds introduced by the SC-Layer as either the number of properties, i.e., safe-ordering constraints (α), the number of disjuncts per property (β), the number of classes (m), or the network depth (δ) are varied. In each plot, the respective parameter varies according to the values on the x-axis, and all other parameters take a default value of $\alpha = 4$, $\beta = 4$, $m = 8$, and $\delta = 6$. As the depth of the network varies, the number of neurons in each layer remains fixed at 1,000 neurons. Reported overheads are averaged over 5 trials.

dimension, n, to be 10. Each dataset is parameterized by α, β, and m, and denoted by $\mathcal{D}(\alpha, \beta, m)$; the procedure for generating these datasets is provided in Appendix C.

We use a dense network with six hidden layers of 1,000 neurons each as a baseline, trained on $\mathcal{D}(4, 4, 8)$. Table 1c shows the overhead introduced by applying SC to our baseline network. We see that the average overhead is approximately a quarter of a millisecond per instance, accounting for a 10× increase in prediction time. Figure 1 provides a more complete picture of the overhead as we vary the number of safe-ordering constraints (α), the number of disjuncts per constraint (β), the number of classes (m), or the depth of the network (δ).

We observe that among these parameters, the overhead is sensitive only to the number of classes. This is to be expected, as the complexity of the SC-Layer scales directly with m (see Sect. 3.4), requiring a topological sort of the m elements of the network's output vector. On the other hand, perhaps surprisingly, increasing the complexity of the safety constraints through either additional safe-ordering constraints or larger disjunctive clauses in the ordering constraints had little effect on the overhead. While in the *worst case* the complexity of the SC-Layer is also dependent on these parameters (Sect. 3.4), if FindSatConstraint finds a satisfiable disjunct quickly, it will short-circuit. The average-case complexity of FindSatConstraint is therefore more nuanced, depending to a greater extent on the specifics of the constraints rather than simply their size. Altogether, these observations suggest that the topological sort in SC-Layer tends to account for the majority of the overhead.

Finally, the results in Fig. 1 concur with what we observed in Sect. 5.3; namely that the overhead is independent of the size of the network.

6   Related Work

Static Verification and Repair of Neural Networks. A number of approaches for verification of already-trained neural networks have been presented in recent years. They have focused on verifying safety properties similar to our safe-ordering constraints. Abstract interpretation approaches [15, 39] verify properties that associate polyhedra with pre- and postconditions. Reluplex [21] encodes a network's semantics as a system of constraints, and poses verification as constraint satisfiability. These approaches can encode safe-ordering constraints, which are a special case of polyhedral postconditions, but they do not provide an effective means to construct safe networks. Other verification approaches [19, 44] do not address safe ordering.

Many of the above approaches can provide counterexamples when the network is unsafe, but none of them are capable of repairing the network. A recent repair approach [40] can provably repair neural networks that have piecewise-linear activations with respect to safety specifications expressed using polyhedral pre- and postconditions. In contrast to our transparency guarantee, they rely on heuristics to favor accuracy preservation.

Safe-by-Construction Learning. Recent efforts seek to learn neural networks that are correct by construction. Some approaches [14, 28, 31] modify the learning objective by adding a penalty for unsafe or incorrect behavior, but they do not provide a safety guarantee for the learned network. Balancing accuracy against the modified learning objective is also a concern. In our work we focus on techniques that provide guarantees without requiring external verifiers.

As discussed in Sect. 5.1, ART [29] aims to learn networks that satisfy safety specifications by updating the loss function used in training. Learning is not guaranteed to converge to zero correctness loss, and the resulting model may not be as accurate as one trained with conventional methods. In contrast, our program transformer is guaranteed to produce a safe network that preserves accuracy.

A similar approach is presented in [33] to enforce local robustness for all input samples in the training dataset. This technique also updates the learning objective and uses a differentiable abstract interpreter for over-approximating the set of reachable outputs. For both this approach and that of [29], the run time of the differentiable abstract interpreter depends heavily on the size and complexity of the network, and it may be difficult or expensive to scale them to realistic architectures.

An alternative way to achieve correct-by-construction learning is to modify the architecture of the neural model. This approach has been employed to construct networks that have a fixed Lipschitz constant [4,27,43], a relational property that is useful for certifying local robustness and ensuring good training behavior. Recent work [25,26] shows how to construct models that achieve relaxed notions of global robustness, where the network is allowed to selectively abstain from prediction at inputs where local robustness cannot be certified. [10] use optimization layers to enforce stability properties of neural network controllers. These techniques are closest to ours in spirit, although we focus on safety specifications, and more specifically safe-ordering constraints, which have not been addressed previously in the literature.

Shielding Control Systems. Recent approaches have proposed ensuring safety of control systems by constructing run-time check-and-correct mechanisms, also referred to as *shields* [2,6,46]. Shields check at run time if the system is headed towards an unsafe state and provide corrections for potentially unsafe actions when necessary. To conduct these run-time checks, shields need access to a model of the environment that describes the environment dynamics, i.e., the effect of controller actions on environment states. Though shields and our proposed SC-Layer share the run-time check-and-correct philosophy, they are designed for different problem settings.

Recovering from Program Errors. Embedding run-time checks into a program to ensure safety is a familiar technique in the program verification literature. Contract checking [13,32], run-time verification [17], and dynamic type checking are all instances of such run-time checks. If a run-time check fails, the program terminates before violating the property. A large body of work also exists on gracefully recovering from errors caused by software issues such as divide-by-zero, null-dereference, memory corruption, and divergent loops [5,22,30,36–38]. These approaches are particularly relevant in the context of long-running programs, when aiming to repair state just enough so that computation can continue.

7   Conclusion and Future Directions

We presented a method for transforming a neural network into a safe-by-construction *self-correcting network*, termed SC-Net, without harming the accuracy of the original network. This serves as a practical tool for providing safety with respect to a broad class of safety specifications, namely, *safe-ordering constraints*, that we characterize in this work.

Unlike prior approaches, our technique guarantees safety without further training or modifications to the network's parameters. Furthermore, the scalability of our approach is not limited by the size or architecture of the model being repaired. This allows it to be applied to large, state-of-the-art models, which is impractical for most other existing approaches.

A potential downside to our approach is the run-time overhead introduced by the SC-Layer. We demonstrate in our evaluation that our approach maintains small overheads (less than one millisecond per instance), due to our vectorized implementation, which leverages GPUs for large-scale parallelism.

In future work, we plan to leverage the differentiability of the SC-Layer to further explore training against the repairs made by the SC-Layer, as this can potentially lead to both accuracy and safety improvements.

Acknowledgment. This material is based upon work supported by the Software Engineering Institute under its FFRDC Contract No. FA8702-15-D-0002 with the U.S. Department of Defense, the National Science Foundation under Grant No. CNS-1943016, DARPA GARD Contract HR00112020006, and the Alfred P. Sloan Foundation.

A   Appendix 1: Proofs

Theorem 2 (Accuracy Preservation). *Given a neural network, $f : \mathbb{R}^n \to \mathbb{R}^m$, and set of constraints, Φ, let $f^{\Phi} := SC^{\Phi}(f)$ and let $F^O : \mathbb{R}^n \to [m]$ be the oracle classifier. Assume that SC satisfies transparency. Further, assume that accuracy is consistent with safety, i.e.,*

$$\forall x \in \mathbb{R}^n \,.\, \exists y \,.\, \Phi(x,y) \,\wedge\, \underset{i}{\mathrm{argmax}}\{y_i\} = F^O(x).$$

Then,

$$\forall x \in \mathbb{R}^n \,.\, F(x) = F^O(x) \implies F^{\Phi}(x) = F^O(x)$$

Proof. Let $x \in \mathbb{R}^n$ such that $F(x) = F^O(x)$. By hypothesis, we have that $\exists y \,.\, \Phi(x,y) \wedge \mathrm{argmax}_i\{y_i\} = F^O(x)$, hence we can apply Property 1 to conclude that $F^{\Phi}(x) = F(x) = F^O(x)$.

We now prove that the transformer presented in Algorithm 3.1, SC, is indeed self-correcting; i.e., it satisfies Properties 2(i) and 2(ii). Recall that this means that f^{Φ} will either return safe outputs vectors, or in the event that Φ is inconsistent at a point, and *only* in that event, return $\bot$.

Let $x : \mathbb{R}^n$ be an arbitrary vector. If $\Phi(x, f(x))$ is initially satisfied, the SC-Layer does not modify the original output $y = f(x)$, and Properties 2(i) and 2(ii) are trivially satisfied. If $\Phi(x, f(x))$ does not hold, we will rely on two key properties of FindSatConstraint and Correct to establish that SC is self-correcting. The first, Property 7, requires that FindSatConstraint either return $\bot$, or else return ordering constraints that are sufficient to establish Φ.

Property 7 (FindSatConstraint). *Let Φ be a set of safe-ordering constraints, $x : \mathbb{R}^n$ and $y : \mathbb{R}^m$ two vectors.*
Then $q = \mathrm{FindSatConstraint}(\Phi, x, y)$ satisfies the following properties:

(i) $q = \bot \iff \forall y' \,.\, \neg\Phi(x, y')$
(ii) $q \neq \bot \implies (\forall y' \,.\, q(y') \implies \Phi(x, y'))$

Proof. The first observation is that the list of ordering constraints in $Q_p := \mathrm{Prioritize}(Q_x, y)$ accurately models the initial set of safety constraints Φ, i.e.,

$$\forall y' \,.\, \Phi(x, y') \iff (\exists q \in Q_p \,.\, q(y')) \tag{5}$$

This stems from the definition of the disjunctive normal form, and from the fact that `Prioritize` only performs a permutation of the disjuncts.

We also rely on the following loop invariant, stating that all disjuncts considered so far, when iterating over `Prioritize`(Q_x, y), were unsatisfiable:

$$\forall q \in Q_p . \; \mathtt{idx}(q, Q_p) < \mathtt{idx}(q_i, Q_p) \implies (\forall y . \neg q(y)) \tag{6}$$

Here, $\mathtt{idx}(q, Q_p)$ returns the index of constraint q in the list Q_p. This invariant is trivially true when entering the loop, since the current q_i is the first element of the list. Its preservation relies on `IsSat`(q) correctly determining whether q is satisfiable, i.e., $\mathtt{IsSat}(q) \iff \exists y . \; q(y)$ [35].

Combining these two facts, we can now establish that `FindSatConstraint` satisfies 7(i) and 7(ii). By definition, `FindSatConstraint`(Φ, x, y) outputs $\bot$ if and only if it traverses the entire list Q_p, never returning a q_i. From loop invariant 6, this is equivalent to $\forall q \in Q_p . \forall y' . \; \neg q(y')$, which finally yields Property 7(i) from Eq. 5. Conversely, if `FindSatConstraint`(Φ, x, y) outputs $q \neq \bot$, then $q \in Q_p$. We directly obtain Property 7(ii) as, for any $y' : \mathbb{R}^m$, $q(y')$ implies that $\Phi(x, y')$ by application of Eq. 5

Next, Property 8 states that `Correct` correctly permutes the output of the network to satisfy the constraint that it is given. Combined with Property 7, this is sufficient to show that `SC` is a self-correcting transformer (Theorem 5).

Property 8 (Correct). *Let q be a satisfiable ordering constraint, and $y : \mathbb{R}^m$ a vector. Then `Correct`(q, y) satisfies q.*

Proof. Let $y_i < y_j$ be an atom in q. Reusing notation from Algorithm 3.3, let $y' = \mathtt{Correct}(q, y)$, $y^s := \mathtt{SortDescending}(y)$, and $\pi := \mathtt{TopologicalSort}(\mathtt{OrderGraph}(q), y)$. We have that (j, i) is an edge in `OrderGraph`(q), which implies that $\pi(j) < \pi(i)$ by Eq. 1. Because the elements of y are sorted in descending order, and assumed to be distinct (Definition 1), we obtain that $y^s_{\pi(i)} < y^s_{\pi(j)}$, i.e., that $y'_i < y'_j$.

Theorem 5 (`SC` is a self-correcting transformer). `SC` *(Algorithm 3.1) satisfies conditions* (i) *and* (ii) *of Definition 2.*

Proof. By definition of Algorithm 3.1, `FindSatConstraint`$(\Phi, x, y) = \bot$ if and only if $f^\Phi(x) = \mathtt{SC}(\Phi)(f)(x)$ outputs $\bot$. We derive from Property 7(i) that this is equivalent to $\forall y' . \; \neg \Phi(x, y')$, which corresponds exactly to Property 2(ii). Conversely, if Φ is satisfiable for input x, i.e., $\exists y' . \; \Phi(x, y')$, then `FindSatConstraint`$(\Phi, x, y)$ outputs $q \neq \bot$. By definition, we have $f^\Phi(x) = \mathtt{Correct}(q, y)$, which satisfies q by application of Property 8, which in turn implies that $\Phi(x, f^\Phi(x))$ by application of Property 7(ii).

Now that we have demonstrated that our approach produces safe-by-construction networks, we next prove that it also preserves the top predicted class when possible, i.e., that SC satisfies *transparency*, as formalized in Property 1.

Let $x : \mathbb{R}^n$ be an arbitrary vector. As in the previous section, if $\Phi(x, f(x))$ is initially satisfied, transparency trivially holds, as the correction layer does not modify the original output $f(x)$. When $\Phi(x, f(x))$ does not hold, we will rely on several additional properties about FindSatConstraint, Correct, and OrderGraph. The first, Property 9, states that whenever the index of the network's top prediction is a root of the graph encoding of q used by FindSatConstraint and Correct, then there exists an output which satisfies q that preserves that top prediction.

Property 9 (OrderGraph). *Let q be a satisfiable, disjunction-free ordering constraint, and $y : \mathbb{R}^m$ a vector. Then,*

$$\underset{i}{\arg\max}\{y_i\} \in Roots(\text{OrderGraph}(q)) \iff$$
$$\exists y'.\ q(y') \ \wedge\ \underset{i}{\arg\max}\{y_i\} = \underset{i}{\arg\max}\{y_i'\}$$

The intuition behind this property is that $i^* := \arg\max_i\{y_i\}$ belongs to the roots of OrderGraph(q) if and only if there is no $y_{i^*} < y_j$ constraint in q; hence since q is satisfiable, we can always permute indices in a solution y' to have $\arg\max_i\{y_i'\} = i^*$. Formally, Lemma 1 in Sect. B.1 entails this property, as it shows that the permutation returned by TopologicalSort satisfies it.

Next, Property 10 formalizes the requirement that whenever FindSatConstraint returns a constraint (rather than $\perp$), then that constraint will not eliminate any top-prediction-preserving solutions that would otherwise have been compatible with the full set of safe-ordering constraints Φ.

Property 10 (FindSatConstraint). *Let Φ be a set of safe-ordering constraints, $x : \mathbb{R}^n$ and $y : \mathbb{R}^m$ two vectors, and $q = \text{FindSatConstraint}(\Phi, x, y)$. Then,*

$$q \neq \perp \ \wedge\ \left(\exists y'.\ \Phi(x, y') \ \wedge\ \underset{i}{\arg\max}\{y_i\} = \underset{i}{\arg\max}\{y_i'\}\right) \implies$$
$$\exists y'.\ q(y') \ \wedge\ \underset{i}{\arg\max}\{y_i\} = \underset{i}{\arg\max}\{y_i'\}$$

Proof. Let us assume that $q \neq \perp$, and that $\exists y'.\ \Phi(x, y') \wedge \arg\max_i\{y_i\} = \arg\max_i\{y_i'\}$. We will proceed by contradiction, assuming that there does not exist y'' such that $q(y'')$ and $\arg\max_i\{y_i\} = \arg\max_i\{y_i''\}$, which entails that $\arg\max_i\{y_i\} \notin Roots(\text{OrderGraph}(q))$ by application of Property 9. In combination with the specification of Prioritize (Property 3), this implies that any

$q' \in Q_p$ such that $\exists y'.\ q'(y') \wedge \operatorname{argmax}_i\{y_i\} = \operatorname{argmax}_i\{y_i'\}$ occurs before q in $\texttt{Prioritize}(Q_x, y)$, i.e., $\operatorname{idx}(q', Q_p) < \operatorname{idx}(q, Q_p)$. From loop invariant 6, we therefore conclude that there does not exist such a $q' \in Q_p$, which contradicts the hypothesis $\Phi(x, y')$ by application of Eq. 5.

Lastly, Property 11 states that $\texttt{Correct}$ (Algorithm 3.3) will always find an output that preserves the original top prediction, whenever the constraint returned by $\texttt{FindSatConstraint}$ allows it. This is the final piece needed to prove Theorem 6, the desired result about the self-correcting transformer.

Property 11 (Correct). *Let q be a satisfiable term, and $y : \mathbb{R}^m$ a vector. Then,*

$$(\exists y'.\ q(y') \wedge \operatorname{argmax}_i\{y_i\} = \operatorname{argmax}_i\{y_i'\})$$

$$\implies \operatorname{argmax}_i\{\texttt{Correct}(q, y)_i\} = \operatorname{argmax}_i\{y_i\}$$

Proof. Assume that there exists y' such that $q(y')$ and $\operatorname{argmax}_i\{y_i\} = \operatorname{argmax}_i\{y_i'\}$. This entails that $\operatorname{argmax}_i(y_i) \in \operatorname{Roots}(\texttt{OrderGraph}(q))$ (Property 9), which in turn implies that $\pi(\operatorname{argmax}_i\{y_i\})$ is 0 (Property 4). By definition of a descending sort, we have that $\operatorname{argmax}_i\{\texttt{Correct}(q, y)_i\} = j$, such that $\pi(j) = 0$, hence concluding that $j = \operatorname{argmax}_i\{y_i\}$ by injectivity of π.

Theorem 6 (Transparency of $\texttt{SC}$). $\texttt{SC}$, *the self-correcting transformer described in Algorithm 3.1 satisfies Property 1.*

Proof. That the $\texttt{SC}$ transformer satisfies transparency is straightforward given Properties 9–11. Let us assume that there exists y' such that $\Phi(x, y')$ and $\operatorname{argmax}_i\{y_i'\} = F(x)$. By application of Property 7(i), this implies that $\texttt{FindSatConstraint}(\Phi, x, f(x))$ outputs $q \neq \bot$, and therefore that there exists y' such that $q(y')$ and $\operatorname{argmax}\{y_i'\} = F(x)$ by application of Property 10, since $F(x)$ is defined as $\operatorname{argmax}_i\{f_i(x)\}$. Composing this fact with Property 11, we obtain that $F^\Phi(x) = F(x)$, since $F^\Phi(x) = \operatorname{argmax}_i\{f_i^\Phi(x)\}$ by definition.

B Appendix 2: Vectorizing Self-Correction

Several of the subroutines of $\texttt{FindSatConstraint}$ and $\texttt{Correct}$ (Algorithms 3.2 and 3.3 presented in Sect. 3) operate on an $\texttt{OrderGraph}$, which represents a conjunction of ordering literals, q. An $\texttt{OrderGraph}$ contains a vertex set, V, and edge set, E, where V contains a vertex, i, for each class in $\{0, \ldots, m-1\}$, and E

Algorithm B.1: Stable Topological Sort

Inputs: A graph, G, represented as an $m \times m$ adjacency matrix, and a vector,
$y : \mathbb{R}^m$
Result: A permutation, $\pi : [m] \to [m]$

1 TopologicalSort(G , y):

2     $P := \texttt{all\_pairs\_longest\_paths}(G)$

3     $\forall\, i,j \in [m]\;.\; P'_{ij} := \begin{cases} y_i & \text{if } P_{ij} \geq 0 \\ \infty & \text{otherwise} \end{cases}$

4     $\forall\, j \in [m]\;.\; v_j := \min_i \{\, P'_{ij} \,\}$
   `// set the value of each vertex to the`
                                `// smallest value among its ancestors`

5     $\forall\, j \in [m]\;.\; d_j := \max_i \{\, P_{ij} \,\}$
   `// calculate the depth of each vertex`

6   **return** $\texttt{argsort}([\; \forall j \in [m]\;.\;(-v_j, d_j)\;])$
   `// break ties in favor of minimum depth`

---

contains an edge, (i, j), from vertex i to vertex j if the literal $y_j < y_i$ is in q. We represent an `OrderGraph` as an $m \times m$ adjacency matrix, M, defined according to Eq. 7.

$$M_{ij} := \begin{cases} 1 & \text{if } (i,j) \in E; \text{ i.e., } y_j < y_i \in q \\ 0 & \text{otherwise} \end{cases} \tag{7}$$

Section B.1 describes the matrix-based algorithm that we use to conduct the stable topological sort that `Correct` (Algorithm 3.3) depends on. It is based on a classic parallel algorithm due to [9], which we modify to ensure that SC satisfies transparency (Property 1). Section B.2 describes our approach to cycle detection, which is able to share much of its work with the topological sort. Finally, Sect. B.3 discusses efficiently prioritizing ordering constraints, needed to ensure that SC satisfies transparency.

B.1 Stable Topological Sort

Our approach builds on a parallel topological sort algorithm given by [9], which is based on constructing an *all pairs longest paths* (APLP) matrix. However, this algorithm is not *stable* in the sense that the resulting order depends only on the graph, and not on the original order of the sequence, even when multiple orderings are possible. While for our purposes this is sufficient for ensuring safety, it is not for transparency. We begin with background on constructing the APLP matrix, showing that it is compatible with a vectorized implementation, and then describe how it is used to perform a stable topological sort.

All Pairs Longest Paths. The primary foundation underpinning many of the graph algorithms in this section is the *all pairs longest paths* (APLP) matrix, which we will denote by P. On acyclic graphs, P_{ij} for $i, j \in [m]$ is defined to be the length of the *longest* path from vertex i to vertex j. Absent the presence of cycles, the distance from a vertex to itself, P_{ii}, is defined to be 0. For vertices i and j for which there is no path from i to j, we let $P_{ij} = -\infty$.

We compute P from M using a matrix-based algorithm from [9], which requires taking $O(\log m)$ matrix *max-distance products*, where the max-distance product is equivalent to a matrix multiplication where element-wise multiplications have been replaced by additions and element-wise additions have been replaced by the pairwise maximum. That is, a matrix product can be abstractly written with respect to operations $\otimes$ and $\oplus$ according to Eq. 8, and the max-distance product corresponds to the case where $x \otimes y := x + y$ and $x \oplus y := \max\{x, y\}$.

$$(AB)_{ij} := (A_{i1} \otimes B_{1j}) \oplus \ldots \oplus (A_{ik} \otimes B_{kj}) \tag{8}$$

Using this matrix product, $P = P^{2^{\lceil \log_2(m) \rceil}}$ can be computed recursively from M by performing a fast matrix exponentiation, as described in Eq. 9.

$$P^k = P^{k/2} P^{k/2} \qquad P^1_{ij} = \begin{cases} 1 & \text{if } M_{ij} = 1 \\ 0 & \text{if } M_{ij} = 0 \ \wedge \ i = j \\ -\infty & \text{otherwise} \end{cases} \tag{9}$$

Stable Sort. We propose a stable variant of the [9] topological sort, shown in Algorithm B.1. Crucially, this variant satisfies Property 4 (Lemma 1), which Sect. 3.2 identifies as sufficient for ensuring transparency. Essentially, the value of each logit y_j is adjusted so that it is at least as small as the smallest logit value corresponding to vertices that are parents of vertex j, including j itself. A vertex, i, is a parent of vertex j if $P_{ij} \geq 0$, meaning that there is some path from vertex i to vertex j or $i = j$. The logits are then sorted in descending order, with ties being broken in favor of minimum depth in the dependency graph. The depth of vertex j is the maximum of the jth column of P_{ij}, i.e., the length of the longest path from any vertex to j. An example trace of Algorithm B.1 is given in Fig. 2. By adjusting y_j into v_j such that for all ancestors, i, of j, $v_i \geq v_j$, we ensure each child vertex appears after each of its parents in the returned ordering–once ties have been broken by depth—as the child's depth will always be strictly larger than that of any of its parents since a path of length d to an immediate parent of vertex j implies the existence of a path of length $d + 1$ to vertex j.

Lemma 1. `TopologicalSort` *satisfies Property 4.*

Proof. Note that the adjusted logit values, v, are chosen according to Eq. 10.

$$v_j := \min_{i \ | \ i \text{ is an ancestor of } j \ \vee \ i=j} \left\{ y_i \right\} \tag{10}$$

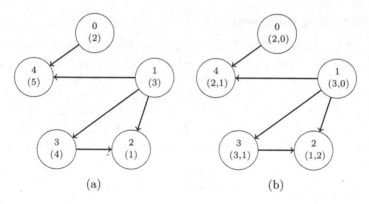

(a)   (b)

Fig. 2. Example trace of Algorithm B.1. **(a)**: The dependency graph and original logit values, y. The values of each logit are provided; the non-bracketed number indicates the logit index and the number in brackets is the logit value, e.g., $y_0 = 2$. Arrows indicate a directed edge in the dependency graph; e.g., we require $y_4 < y_0$. **(b)**: updated values passed into `argsort` as a tuple. For example, y_4 is assigned $(2, 1)$, as its smallest ancestor (y_0) has logit value 2 in (a) and its depth is 1; and y_2 is assigned value $(1, 2)$ because its logit value in (a), 1, is already smaller than that any of its parents, and its depth is 2. The values are sorted by *decreasing* value and *increasing* depth, thus the final order is $\langle y_1, y_3, y_0, y_4, y_2 \rangle$, corresponding to the permutation π, where $\pi(0) = 2$, $\pi(1) = 0$, $\pi(2) = 4$, $\pi(3) = 1$, and $\pi(4) = 3$.

We observe that *(i)* for all root vertices, i, $v_i = y_i$, and *(ii)* the root vertex with the highest original logit value will appear first in the topological ordering. The former follows from the fact that the root vertices have no ancestors. The latter subsequently follows from the fact that the first element in a valid topological ordering must correspond to a root vertex. Thus if $\text{argmax}_i\{y_i\} = i^* \in \text{Roots}(g)$, then i^* is the vertex with the highest logit value, and so by *(ii)*, it will appear first in the topological ordering produced by `TopologicalSort`, establishing Property 4.

B.2 Cycle Detection

`IsSat`, a subroutine of `FindSatConstraint` (Algorithm 3.2) checks to see if an ordering constraint, q, is satisfiable by looking for any cycles in the corresponding dependency graph, `OrderGraph`(q). Here we observe that the existence of a cycle can easily be decided from examining P, by checking if $P_{ii} > 0$ for some $i \in [m]$; i.e., if there exists a non-zero-length path from any vertex to itself. Since $P_{ii} \geq 0$, this is equivalent to $\text{Trace}(P) > 0$. While strictly speaking, P_{ij}, as constructed by [9], only reflects the longest path from i to j in *acyclic* graphs, it can nonetheless be used to detect cycles in this way, as for any $k \leq m$, P_{ij} is guaranteed to be at least k if there exists a path of length k from i to j, and any cycle will have length at most m.

B.3   Prioritizing Root Vertices

As specified in Property 3, in order to satisfy transparency, the search for a satisfiable ordering constraint performed by `FindSatConstraint` must prioritize constraints, q, in which the original predicted class, $F(x)$, is a root vertex in q's corresponding dependency graph. We observe that root vertices can be easily identified using the dependency matrix M. The in-degree, d_j^{in}, of vertex j is simply the sum of the jth column of M, given by Eq. 11. Meanwhile, the root vertices are precisely those vertices with no ancestors, that is, those vertices j satisfying Eq. 11.

$$d_j^{in} = \sum_{i \in [m]} M_{ij} = 0 \tag{11}$$

In the context of `FindSatConstraint`, the subroutine `Prioritize` lists ordering constraints q for which $d_{F(x)}^{in} = 0$ in `OrderGraph`(q) before any other ordering constraints. To save memory, we do not explicitly list and sort all the disjuncts of Q_x (the DNF form of the active postconditions for x); rather we iterate through them one at a time. This can be done by, e.g., iterating through each disjunct twice, initially skipping any disjunct in which $F(x)$ is not a root vertex, and subsequently skipping those in which $F(x)$ is a root vertex.

C   Appendix 3: Generation of Synthetic Data

In Sect. 5.4, we utilize a family of synthetic datasets with associated safe-ordering constraints that are randomly generated according to several specified parameters, allowing us to assess how aspects such as the number of constraints (α), the number of disjunctions per constraint (β), and the dimension of the output vector (m) impact the run-time overhead. In our experiments, we fix the input dimension, n, to be 10. The synthetic data, which we will denote by $\mathcal{D}(\alpha, \beta, m)$, are generated according to the following procedure.

(i) First, we generate α random safe-ordering constraints. The preconditions take the form $b_\ell \leq x \leq b_u$, where b_ℓ is drawn uniformly at random from $[0.0, 1.0 - \epsilon]$ and $b_u := b_\ell + \epsilon$. We choose $\epsilon = 0.4$ in our experiments; as a result, the probability that any two preconditions overlap is approximately 30%. The ordering constraints are disjunctions of β randomly-generated cycle-free ordering graphs of m vertices, i.e., β disjuncts. Specifically, in each graph, we include each edge, (i, j), for $i \neq j$ with equal probability, and require further that at least one edge is included, and the expected number of edges is γ (we use $\gamma = 3$ in all of our experiments). Graphs with cycles are resampled until a graph with no cycles is drawn.

(ii) Next, for each safe-ordering constraint, ϕ, we sample N/α random inputs, x, uniformly from the range specified by the precondition of ϕ. In all of our experiments we let $N = 2{,}000$. For each x, we select a random disjunct from the postcondition of ϕ, and find the roots of the corresponding ordering graph. We select a label, y^* for x uniformly at random from this set of roots, i.e., we pick a random label for each point that is consistent with the property for that point.

(iii) Finally, we generate N random points that do not satisfy any of the preconditions of the α safe-ordering constraints. We label these points via a classifier trained on the N labeled points already generated in (ii). This results in a dataset of $2N$ labeled points, where 50% of the points are captured by at least one safe-ordering constraint.

References

1. Abadi, M., et al.: TensorFlow: a system for large-scale machine learning. In: 12th USENIX Symposium on Operating Systems Design and Implementation (OSDI 2016), pp. 265–283 (2016)
2. Alshiekh, M., Bloem, R., Ehlers, R., Könighofer, B., Niekum, S., Topcu, U.: Safe reinforcement learning via shielding. In: Thirty-Second AAAI Conference on Artificial Intelligence (2018)
3. Anderson, G., Pailoor, S., Dillig, I., Chaudhuri, S.: Optimization and abstraction: a synergistic approach for analyzing neural network robustness. In: Proceedings of the 40th ACM SIGPLAN Conference on Programming Language Design and Implementation, PLDI 2019, pp. 731–744. Association for Computing Machinery, New York (2019). https://doi.org/10.1145/3314221.3314614
4. Anil, C., Lucas, J., Grosse, R.: Sorting out Lipschitz function approximation. In: International Conference on Machine Learning, pp. 291–301. PMLR (2019)
5. Berger, E.D., Zorn, B.G.: DieHard: probabilistic memory safety for unsafe languages. In: Proceedings of the 27th ACM SIGPLAN Conference on Programming Language Design and Implementation, PLDI 2006, pp. 158–168. Association for Computing Machinery, New York (2006)
6. Bloem, R., Könighofer, B., Könighofer, R., Wang, C.: Shield synthesis: In: Baier, C., Tinelli, C. (eds.) TACAS 2015. LNCS, vol. 9035, pp. 533–548. Springer, Heidelberg (2015). https://doi.org/10.1007/978-3-662-46681-0_51
7. Brown, T.B., et al.: Language models are few-shot learners. arXiv preprint arXiv:2005.14165 (2020)
8. Clarkson, M.R., Schneider, F.B.: Hyperproperties. In: 2008 21st IEEE Computer Security Foundations Symposium, pp. 51–65 (2008). https://doi.org/10.1109/CSF.2008.7
9. Dekel, E., Nassimi, D., Sahni, S.: Parallel matrix and graph algorithms. SIAM J. Comput. 10, 657–675 (1981)
10. Donti, P.L., Roderick, M., Fazlyab, M., Kolter, J.Z.: Enforcing robust control guarantees within neural network policies. In: International Conference on Learning Representations (2021). https://openreview.net/forum?id=5lhWG3Hj2By
11. Dvijotham, K., Stanforth, R., Gowal, S., Mann, T., Kohli, P.: A dual approach to scalable verification of deep networks. In: Proceedings of the Thirty-Fourth Conference Annual Conference on Uncertainty in Artificial Intelligence (UAI 2018), Corvallis, Oregon, pp. 162–171. AUAI Press (2018)
12. Ehlers, R.: Formal verification of piece-wise linear feed-forward neural networks. In: D'Souza, D., Narayan Kumar, K. (eds.) ATVA 2017. LNCS, vol. 10482, pp. 269–286. Springer, Cham (2017). https://doi.org/10.1007/978-3-319-68167-2_19

13. Findler, R.B., Felleisen, M.: Contracts for higher-order functions. In: Proceedings of the Seventh ACM SIGPLAN International Conference on Functional Programming, ICFP 2002 (2002)
14. Fischer, M., Balunovic, M., Drachsler-Cohen, D., Gehr, T., Zhang, C., Vechev, M.: DL2: training and querying neural networks with logic. In: International Conference on Machine Learning, pp. 1931–1941. PMLR (2019)
15. Gehr, T., Mirman, M., Drachsler-Cohen, D., Tsankov, P., Chaudhuri, S., Vechev, M.: AI2: Safety and robustness certification of neural networks with abstract interpretation. In: 2018 IEEE Symposium on Security and Privacy (SP), pp. 3–18 (2018)
16. Guttmann, W., Maucher, M.: Variations on an ordering theme with constraints. In: Navarro, G., Bertossi, L., Kohayakawa, Y. (eds.) TCS 2006. IIFIP, vol. 209, pp. 77–90. Springer, Boston, MA (2006). https://doi.org/10.1007/978-0-387-34735-6_10
17. Havelund, K., Rosu, G.: Monitoring programs using rewriting. In: Proceedings 16th Annual International Conference on Automated Software Engineering (ASE 2001), pp. 135–143 (2001)
18. He, K., Zhang, X., Ren, S., Sun, J.: Deep residual learning for image recognition. In: IEEE Conference on Computer Vision and Pattern Recognition (CVPR) (2016)
19. Huang, X., Kwiatkowska, M., Wang, S., Wu, M.: Safety verification of deep neural networks. In: Majumdar, R., Kunčak, V. (eds.) CAV 2017. LNCS, vol. 10426, pp. 3–29. Springer, Cham (2017). https://doi.org/10.1007/978-3-319-63387-9_1
20. Julian, K.D., Kochenderfer, M.J., Owen, M.P.: Deep neural network compression for aircraft collision avoidance systems. J. Guid. Control Dyn. **42**(3), 598–608 (2019)
21. Katz, G., Barrett, C., Dill, D.L., Julian, K., Kochenderfer, M.J.: Reluplex: an efficient SMT solver for verifying deep neural networks. In: Majumdar, R., Kunčak, V. (eds.) CAV 2017. LNCS, vol. 10426, pp. 97–117. Springer, Cham (2017). https://doi.org/10.1007/978-3-319-63387-9_5
22. Kling, M., Misailovic, S., Carbin, M., Rinard, M.: Bolt: on-demand infinite loop escape in unmodified binaries. In: Proceedings of the ACM International Conference on Object Oriented Programming Systems Languages and Applications, OOPSLA 2012, pp. 431–450. Association for Computing Machinery, New York (2012)
23. Kochenderfer, M.J., et al.: Optimized airborne collision avoidance, pp. 249–276 (2015)
24. Krizhevsky, A., Hinton, G.: Learning multiple layers of features from tiny images. Technical report 0, University of Toronto, Toronto, Ontario (2009)
25. Leino, K., Fredrikson, M.: Relaxing local robustness. In: Advances in Neural Information Processing Systems (NeurIPS) (2021)
26. Leino, K., Wang, Z., Fredrikson, M.: Globally-robust neural networks. In: International Conference on Machine Learning (ICML) (2021)
27. Li, Q., Haque, S., Anil, C., Lucas, J., Grosse, R.B., Jacobsen, J.H.: Preventing gradient attenuation in lipschitz constrained convolutional networks. In: Advances in Neural Information Processing Systems 32, pp. 15390–15402 (2019)
28. Li, T., Gupta, V., Mehta, M., Srikumar, V.: A logic-driven framework for consistency of neural models. In: Inui, K., Jiang, J., Ng, V., Wan, X. (eds.) Proceedings of the 2019 Conference on Empirical Methods in Natural Language Processing and the 9th International Joint Conference on Natural Language Processing, EMNLP-IJCNLP 2019, Hong Kong, China, 3–7 November 2019, pp. 3922–3933. Association for Computational Linguistics (2019). https://doi.org/10.18653/v1/D19-1405

29. Lin, X., Zhu, H., Samanta, R., Jagannathan, S.: ART: abstraction refinement-guided training for provably correct neural networks. In: 2020 Formal Methods in Computer Aided Design (FMCAD), pp. 148–157 (2020)

30. Long, F., Sidiroglou-Douskos, S., Rinard, M.: Automatic runtime error repair and containment via recovery shepherding. In: Proceedings of the 35th ACM SIGPLAN Conference on Programming Language Design and Implementation, PLDI 2014, pp. 227–238. Association for Computing Machinery, New York (2014)

31. Madry, A., Makelov, A., Schmidt, L., Tsipras, D., Vladu, A.: Towards deep learning models resistant to adversarial attacks. In: International Conference on Learning Representations (2018). https://openreview.net/forum?id=rJzIBfZAb

32. Meyer, B.: Eiffel: The Language (1992)

33. Mirman, M., Gehr, T., Vechev, M.: Differentiable abstract interpretation for provably robust neural networks. In: International Conference on Machine Learning, pp. 3578–3586. PMLR (2018)

34. Müller, C., Serre, F., Singh, G., Püschel, M., Vechev, M.: Scaling polyhedral neural network verification on GPUS. In: Proceedings of Machine Learning and Systems 3 (2021)

35. Nieuwenhuis, R., Rivero, J.M.: Practical algorithms for deciding path ordering constraint satisfaction. Inf. Comput. **178**(2), 422–440 (2002). https://doi.org/10.1006/inco.2002.3146

36. Perkins, J.H., et al.: Automatically patching errors in deployed software. In: Proceedings of the ACM SIGOPS 22nd Symposium on Operating Systems Principles, SOSP 2009, pp. 87–102. Association for Computing Machinery, New York (2009)

37. Qin, F., Tucek, J., Sundaresan, J., Zhou, Y.: Rx: treating bugs as allergies–a safe method to survive software failures. In: Proceedings of the Twentieth ACM Symposium on Operating Systems Principles, SOSP 2005, pp. 235–248. Association for Computing Machinery, New York (2005)

38. Rinard, M., Cadar, C., Dumitran, D., Roy, D.M., Leu, T., Beebee, W.S.: Enhancing server availability and security through failure-oblivious computing. In: Proceedings of the 6th Conference on Symposium on Operating Systems Design and Implementation - Volume 6, OSDI 2004, p. 21. USENIX Association, Berkeley (2004)

39. Singh, G., Gehr, T., Püschel, M., Vechev, M.: An abstract domain for certifying neural networks. In: Proceedings of the ACM on Programming Languages, 3(POPL), January 2019

40. Sotoudeh, M., Thakur, A.V.: Provable repair of deep neural networks. In: Proceedings of the 42nd ACM SIGPLAN International Conference on Programming Language Design and Implementation, pp. 588–603 (2021)

41. Tan, M., Le, Q.: EfficientNet: rethinking model scaling for convolutional neural networks. In: International Conference on Machine Learning, pp. 6105–6114. PMLR (2019)

42. Tan, M., Le, Q.V.: EfficientNetV2: smaller models and faster training. arXiv preprint arXiv:2104.00298 (2021)

43. Trockman, A., Kolter, J.Z.: Orthogonalizing convolutional layers with the Cayley transform. In: International Conference on Learning Representations (2021)

44. Urban, C., Christakis, M., Wüstholz, V., Zhang, F.: Perfectly parallel fairness certification of neural networks. In: Proceedings of the ACM on Programming Languages, 4(OOPSLA), November 2020. https://doi.org/10.1145/3428253

45. Wu, H., et al.: Parallelization techniques for verifying neural networks. In: 2020 Formal Methods in Computer Aided Design, FMCAD 2020, Haifa, Israel, 21–24 September 2020, pp. 128–137. IEEE (2020). https://doi.org/10.34727/2020/isbn. 978-3-85448-042-6_20

46. Zhu, H., Xiong, Z., Magill, S., Jagannathan, S.: An inductive synthesis framework for verifiable reinforcement learning. In: Proceedings of the 40th ACM SIGPLAN Conference on Programming Language Design and Implementation, PLDI 2019, pp. 686–701. Association for Computing Machinery, New York (2019). https://doi. org/10.1145/3314221.3314638

Formal Specification for Learning-Enabled Autonomous Systems

Saddek Bensalem[1], Chih-Hong Cheng[2], Xiaowei Huang[3],
Panagiotis Katsaros[4]($\boxtimes$), Adam Molin[5], Dejan Nickovic[6], and Doron Peled[7]

[1] University Grenoble Alpes, VERIMAG, Grenoble, France
[2] Fraunhofer IKS, Munich, Germany
[3] University of Liverpool, Liverpool L69 3BX, UK
[4] Aristotle University of Thessaloniki, Thessaloniki, Greece
katsaros@csd.auth.gr
[5] DENSO AUTOMOTIVE Deutschland GmbH, Eching, Germany
[6] AIT Austrian Institute of Technology, Vienna, Austria
[7] Bar Ilan University, Ramat Gan, Israel

Abstract. The formal specification provides a uniquely readable description of various aspects of a system, including its temporal behavior. This facilitates testing and sometimes automatic verification of the system against the given specification. We present a logic-based formalism for specifying learning-enabled autonomous systems, which involve components based on neural networks. The formalism is based on first-order past time temporal logic that uses predicates for denoting events. We have applied the formalism successfully to two complex use cases.

Keywords: Learning-enabled systems · Formal specification · Neural networks · First-order LTL

1 Introduction

The application of formal methods to software artefacts requires the use of formal specification. A specification formalism defines the permitted behaviors or the intended architecture of a system in a uniquely readable manner. It can be used as a contract between different project stakeholders, including the customers, designers, developers and quality assurance teams. Common formalisms include temporal logics and various graph structures or state machines. Different formalisms can be combined together to describe different aspects of the system, such as in UML [16]. In addition, some formalisms, such as state-charts, employ visual notation in order to better demonstrate the specification.

The challenge we are undertaking here is to adopt a formalism that can describe systems with, possibly, timing and cyber-physical components, that

Supported by the european project Horizon 2020 research and innovation programme under grant agreement No. 956123.
C.-H. Cheng—The work is primarily conducted during his service at DENSO.

are *learning-enabled*, or, in other words, include components that involve neural networks (NNs), trained using deep learning. NNs have strongly impacted on the computer applications in the last decade, including object recognition and natural language processing. The structure of a NN is quite simple: layers of components called "artificial neurons", which have some numerical values, feeding values to the next layer through some linear transformation and then applying an *activation function*, which is a nonlinear transformation. Specifying systems that include components based on NNs is challenging, since a NN has different characteristics from the usual state-transition model.

For example, classifying objects in a picture is commonly performed using a NN. If one considers the values of the individual neurons and the constants used to calculate the transformation between the layers, then the number of possible states is astronomical; moreover, there is no known direct connection between the states and the results associated with the NN. Then it is hardly reasonable to specify directly the connection between the values of the different components of the NN and the classification result, e.g., identifying the object that appears in the picture as a pedestrian or bicycle rider.

Learning-enabled systems appear nowadays in a growing number of applications. This stems from the ability of NNs to provide new identification capabilities, e.g., related to vision and speech recognition. Such systems are often intended to interact with people and the environment, most notably, autonomous driving. This makes these applications highly safety-critical.

We introduce a specification formalism that is based on abstracting away the internal structure of the NN, including the internal values of the different neurons; instead, inspired by [4], our specification asserts about objects that are represented using the NN and related values they stand for. We aim to an intuitive yet expressive formal language, which will match the specification requirements for learning enabled autonomous systems. The adequacy of our formalism has been tested against different requirements for the use cases in the FOCETA EU2020 project[1].

The rest of the paper is structured as follows. Section 2 introduces the syntax and semantics of the specification language. Section 3 presents representative formal specifications from two learning-enabled autonomous systems from the FOCETA project. Section 4 reviews the related work and finally, we provide our concluding remarks, most notably the rationale behind the proposed formalism.

2 Formal Specifications

2.1 Event-Based Abstraction

Given the difficulty of specifying a system that includes a NN based on its set of states, we propose an *event-based* abstraction that hides the details of the NN structure [4]. The specification is defined over *relations* or *predicates* of the form $p(a_1, \ldots, a_n)$ over domains $\mathcal{D}_1, \ldots, \mathcal{D}_n$, where for $1 \leq i \leq n$, $a_i \in \mathcal{D}_i$ is a value from the domain $\mathcal{D}_i$. These domains can be, e.g., the integers, the

---

[1] http://www.foceta-project.eu/.

reals or strings. One can also use (state-dependent) Boolean variables, which are degenerate predicates with 0 parameters. Formally, let $p \subseteq \mathcal{D}_{i_1} \times \ldots \times \mathcal{D}_{i_m}$ be a relation over subsets of these domains. An *event-based state* or *eb-state* is a set of tuples from these relations. We can restrict each relation to contain *exactly* or *at most one* tuple, or do not restrict them, depending on the type of system that is modeled. An *execution* is a finite or infinite sequence of eb-states. A *trace* is a finite prefix of an execution.

Examples of some conventions one can adopt for modeling of systems include:

- For runtime-verification, each relation consists of at most one tuple. In some cases, there is only one tuple of one relation in a state.
- Representing real-time can be achieved by equiping each state with a single unary relation *time(t)*, where t corresponds to time.
- The output of an object recognition NN can be the following tuples: *object_type(ob)*, *accurracy(pr)*, *bounding_box(x_1, y_1, x_2, y_2)*, where *ob* is the object type, e.g., 'road-sign', 'car'; $0 \leq pr \leq 1$ is the perceived probability of recognition by the NN, and (x_1, y_1), (x_2, y_2) are the bottom left and top right point of the bounding box around the identified object. Object recognition systems can include multiple objects that are recognized in a single frame.
- One can use different units of measurements when referring to time or other physical components, e.g., distance or energy level.

When modeling a system, the assumed conventions, e.g., the number of tuples per relation allowed in a state and the unit of measurements need to be presented separately from the specification formulas.

Syntax. The formulas of the core logic, which is based on a *first-order* extension [17] of the *past* portion of Linear Temporal Logic (LTL) [19] are defined by the following grammar, where a_i is a constant representing a value in some domain $\mathcal{D}$, and x_i denotes a variable over the same domain $domain(x_i)$. The value of x_i must be from the domain associated with this variable.

$$\varphi ::= true \mid false \mid p(x_1, \ldots, x_n) \mid (\varphi \vee \varphi) \mid (\varphi \wedge \varphi) \mid$$
$$\neg \varphi \mid (\varphi \, S \, \varphi) \mid \ominus \varphi \mid \exists x \, \varphi \mid \forall x \, \varphi \mid e \sim e$$

where $\sim \in \{<, \leq, >, \geq, =, \neq\}$ and $e ::= x \mid a \mid e + e \mid e - e \mid e \times e \mid e \, / \, e$. We read $(\varphi S \psi)$ as φ *since* ψ.

Semantics. Let γ be an assignment to the variables that appear free in a formula φ, with $\gamma(x)$ returning the value of the variable x under the assignment γ. Then $(\gamma, \sigma, i) \models \varphi$ means that φ holds for the assignment γ, and the trace σ of length i. We denote the ith event of σ by $\sigma[i]$. Note that by using past operators, the semantics is not affected by states s_j with $j > i$ that appear in longer prefixes than σ of the execution. Let $vars(\varphi)$ be the set of free variables of a subformula φ (i.e., x is not within the scope of a quantifier $\forall$ or $\exists$, as in $\forall x \, \varphi$, $\exists x \, \varphi$). We denote by $\gamma|_{vars(\varphi)}$ the restriction (projection) of an assignment γ to the free variables appearing in φ.

Let $v(e, \gamma)$ be the value assigned to an expression e under the assignment γ:

- $v(a, \gamma) = a$, when a is a constant.
- $v(x, \gamma) = \gamma(v)$, when x is a varaible.
- $v(e_1 + e_2, \gamma) = v(e_1, \gamma) + v(e_2, \gamma)$, and similarly for '$-$', '$\times$' and '$/$'.

Let ϵ be an empty assignment. In any of the following cases, $(\gamma, \sigma, i) \models \varphi$ is defined when γ is an assignment over $vars(\varphi)$, and $i \geq 1$.

- $(\epsilon, \sigma, i) \models true$.
- $(\gamma, \sigma, i) \models p(y_1, \ldots, y_n)$ if $p(v(y_1), \ldots, v(y_n)) \in \sigma[i]$.
- $(\gamma, \sigma, i) \models (\varphi \wedge \psi)$ if $(\gamma|_{vars(\varphi)}, \sigma, i) \models \varphi$ and $(\gamma|_{vars(\psi)}, \sigma, i) \models \psi$.
- $(\gamma, \sigma, i) \models \neg \varphi$ if not $(\gamma, \sigma, i) \models \varphi$.
- $(\gamma, \sigma, i) \models (\varphi\ S\ \psi)$ if for some $1 \leq j \leq i$, $(\gamma|_{vars(\psi)}, \sigma, j) \models \psi$ and for all $j < k \leq i$, $(\gamma|_{vars(\varphi)}, \sigma, k) \models \varphi$.
- $(\gamma, \sigma, i) \models \ominus \varphi$ if $i > 1$ and $(\gamma, \sigma, i - 1) \models \varphi$.
- $(\gamma, \sigma, i) \models \exists x\ \varphi$ if there exists $a \in domain(x)$ such that[2] $(\gamma\,[x \mapsto a], \sigma, i) \models \varphi$.
- $(\gamma, \sigma, i) \models e_1 < e_2$ if $v(e_1, \gamma) < v(e_2, \gamma)$, and similarly for the relations '$\leq$', '$>$', '$\geq$', '$=$' and '$\neq$'.

The rest of the operators are defined as syntactic sugar using the operators defined in the above semantic definitions: $false = \neg true$, $\forall x \varphi = \neg \exists x \neg \varphi$, $(\varphi \vee \psi) = \neg(\neg\varphi \wedge \neg\psi)$. We can also define the following useful operators: $P\varphi = (true\ S\ \varphi)$ (for "Previously") and $H\varphi = (false\ R\ \varphi)$ (for "always in the past").

The specification needs to appear in a context that includes the interpretations of the relations that are used. For clarity, we sometimes denote the domain within the formula during quantification. For example, $\exists x \in Obj\ \varphi$ specifies explicitly the domain name Obj for values of the variable x, which may otherwise be understood from the context where the formula appears.

Intended Interpretation. We restrict ourselves to *safety properties* [1], and the interpretation of a formula φ is over *every prefix* of the execution sequence. To emphasize that the interpretation is over all the prefixes, we can use the **G** modality from future LTL, writing $\mathbf{G}\varphi$, where φ is a first-order past LTL formula. There are several reasons for this. First, and foremost, safety properties are most commonly used; in many cases, a non-safety property, which guarantees some progress "eventually" as in the future LTL operator $\Diamond$ [19], without specifying a distinct time, can be replaced with a fixed deadline; then it becomes a safety property. For example, instead of expressing that every *request* is eventually followed by an *acknowledge*, specifying that any request must be followed by an acknowledge *within no more than* 10*ms* is a safety property. In addition, safety properties are often more susceptible to the application of formal methods; a notable example for our context is the ability to perform runtime verification on linear temporal logic with data with specification formalisms similar to the one used here [7,17].

Dealing with Quantitative Progress of Time. Temporal logic specification often abstracts away from using real time, where the intended model uses discrete

---

[2] $\gamma\,[x \mapsto a]$ is the overriding of γ with the binding $[x \mapsto a]$.

progress between states. We make use of time predicates, in particular, *time* with a single integer or real parameter, which can be part of the events in an execution. For example, the term $time(t1)$ can refer to a timer that reports a value of $t1$. By comparing different values of such terms, one refers to the amount of time elapsed between related events.

Examples of Specifications

- $\forall x \, (closed(x) \rightarrow \ominus(\neg closed(x) S open(x)))$. [For each file, if we closed a file, its the first time we close it since it was opened.]
- $\forall x \, \forall t_1 \, \exists t_2 \, ((t_1 - t_2 < 90 \wedge time(t_1) \wedge closed(x)) \rightarrow \ominus(closed(x)S(open(x) \wedge time(t_2))))$ [Every file, cannot remain opened before it is closed more than 90s. Note that the interpretation of 90 as a meausre of seconds and of $time(t)$ as a predicate that holds if t is the current time is a matter of choice.]
- $\forall t1((time(t1) \wedge \neg stopped(car1)) \rightarrow \neg \exists t2(\neg stopped(car1)S(id(stop\_sign, pr) \wedge time(t2) \wedge pr \geq 0.9 \wedge t1 - t2 > 0.3)))$ [At any time, if *car1* is not stopped, no stop sign has been identified in the last 0.3 s with probability ≥ 0.9.]

3 Use Case Specifications from Learning-Enabled Autonomous Systems

In this section, we will show how the specification formalism we proposed allows describing the properties of two challenging use cases:

1. A safe and secure intelligent automated valet parking (AVP) system. This is an L4 autonomous driving system with a fixed Operational Design Domain (ODD) on a given set of parking lots. A user owning a vehicle equipped with the AVP functionality stops the car in front of the parking lot entry area. Whenever the user triggers the AVP function, the vehicle communicates with the infrastructure and parks the car at designated regions (assigned by the infrastructure). The system is expected to operate under mixed traffic, i.e., the parking lot will have other road users including pedestrians and vehicles.
2. A life-critical anaesthetic drug target control infusion system. This use case concerns with the manipulation of hypnotic sedative drugs, and the ability to provide new and breakthrough technology to cope with a better control of sedation status in the patient. Since each patient is unique, no single dosage of an anesthetic is likely to be appropriate for all patients. In addition, providing an under or over-dosage of an anesthetic is highly undesirable. The development of this autonomous controller would facilitate the work of the anaesthesiologists and increase patient safety through better control of the depth of anesthesia. For this development, the verification and validation of the controller prior to any clinical investigation with real patients is essential, so a virtual testbench platform with a complete test plan is required for this.

3.1 Automated Valet Parking System

Object Detection. Object detection is a key component that aims at recognising objects from sensory input. The sensory input can be understood as a sequence of single inputs such as images. Usually, a deep learning system (such as YOLO [23]) is applied to return a set of detected objects from a single image, although the result may also depend on the detection results of a sequence of images. We consider an object detection component (ODC), for which there are three major classes of specifications:

1. Functional specifications, concerning whether the object detector exhibits the expected behaviour in normal circumstances on a single image frame.
2. Temporal specifications, for the expected behaviour in sequential inputs.
3. Robustness specifications, concerning whether and how the expected behaviour may be affected by perturbations to the input.

Functional Specifications. While there may be various specifications, we consider the following as a typical one:
For every object y in the world, if y is a pedestrian that stands within X meters of range, then the ODC will detect some object z as a pedestrian at almost the same position as y (within ε).
This can be expressed as the following formula:

$$\forall y \in Obj\ ((pedestrian(y) \wedge range(y)) \rightarrow detect(y)) \tag{1}$$

where: Obj is assumed to be the set of all *objects* occupying the world (this refers to the ground truth); $pedestrian(y)$ is a predicate that is true iff y is a pedestrian (this refers to the ground truth); $range(y)$ is defined as $distance\_ego(y) \leq X$ where X is the "X meters" parameter of the English spec, and $distance\_ego(y)$ returns the distance of y from the ego vehicle (this also refers to the ground truth); $detect(y)$ is defined as ϕ_2 as follows

$$\exists z \in ODC\_Obj\ (ODC\_pedestrian(z) \wedge |ODC\_position(z), position(y)| \leq \epsilon)$$

where: ODC_Obj is the set of objects detected by the ODC (i.e., "the system" that this spec refers to), $ODC\_pedestrian(z)$ is a predicate which is true iff z is classified as a pedestrian by ODC; $ODC\_position(z)$ is the position of z as returned by ODC; $|a, b|$ is a function that returns the distance between positions a and b; ϵ is a parameter that represents how "close" two positions are.

Temporal Specifications. While the specification in (1) considers whether the ODC performs correctly in a single frame of the video stream, it is possible that the overall functionality of the ODC may not be compromised by the failure of a single frame. Therefore, we may consider temporal specifications such as

$$\mathbf{G}\phi_1 \tag{2}$$

Besides, we may consider other temporal specifications such as:
In a sequence of images from a video feed, any object to be detected should not be missed more than 1 in X frames.

This property can be formalised with the following formula:

$$\pi := \mathbf{G}(\neg\phi_1 \rightarrow \bigwedge_{t=1}^{X-1} \ominus^t\phi_1). \tag{3}$$

which, intuitively, guarantees that once there is an incorrect detection at time t, the outputs at previous $X - 1$ steps should be all correct.

Robustness Specifications. The aforementioned classes of specifications do not consider the possible perturbations to the input. However, perturbations such as benign/natural noises or adversarial/security attacks can be typical to an ODC, which works with natural data. We consider the following specification:
For any input x, which contains a pedestrian y, the detection will be the same within certain perturbation δ with respect to the distance measure d.
This can be expressed as follows (*Input* is the set of possible image frames):

$$\mathbf{G}\forall x, x' \in Input, \exists y, y' \in ODC\_Obj, ODC\_pedestrian(y) \wedge$$
$$d(x, x') \leq \delta \rightarrow pedestrian(y') \tag{4}$$

Planning and Control. *Planning* refers to the task of making decisions to achieve high-level goals, such as moving the vehicle from a start location (e.g. drop-off space for a parking system) to the goal location, while avoiding obstacles and optimizing over some parameter (e.g. shortest path). *Control* is responsible to execute the actions that have been generated by the higher-level planning tasks and generate the necessary inputs to the autonomous system, in order to realize the desired motions.

Planning is usually further decomposed into *mission planning* and *path* planning. Mission planning represents the highest level decisions regarding the route (sequence of way-points on the map) to be followed, whereas the task of path planning refers to the problem of generating a collision-free trajectory based on the high-level mission plan and the current location of the vehicle.

Mission Planning. The mission planner must ensure that (i) traffic rules are followed (e.g., wait at stop sign) and (ii) obstacles are avoided.
Traffic rule:
At any time, if ego is not stopped, it is not the case that a red_light was sensed within the last second.

$$\mathbf{G}\forall t_1(time(t_1) \wedge \neg stopped(ego)) \rightarrow \tag{5}$$
$$\neg\exists t_2(\neg stopped(ego) \; S \; sensed(red\_light) \wedge time(t_2) \wedge t_1 - t_2 > 1)$$

where $stopped(ego)$ abstracts the respective signal activated by the mission planner and $sensed(red\_light)$ abstracts the output of the perception system.
Collision avoidance:

The planner shall calculate a reference trajectory that keeps a distance d_{ttc} (ttc: time-to-collision) [or d_{safety}] to obstacles, e.g., (1 s) [or (1.0 m).

$$\mathbf{G}\forall(p,v) \in traj_{\mathrm{ref}}, \forall o \in Obj \quad (distance(p,v,position(o)) \geq d_{ttc}) \qquad (6)$$

or

$$\mathbf{G}\forall(p,v) \in traj_{\mathrm{ref}}, \forall o \in Obj \quad (|p,position(o)| \geq d_{safety}) \qquad (7)$$

where p is the position and v is the velocity of the ego vehicle on a waypoint of the reference trajectory $traj_{\mathrm{ref}}$ (a finite set of way point/velocity tuples) and $position(o)$ is the position of object o.

Path Planning involves requirements for the computed path to some (intermediate) goal location that may refer to the current location of the system.
Feasible path to the parking lot:
At time C (some constant), the latest, the parking lot (goal) is reached.

$$\mathbf{G}\neg(time(C) \wedge H\ position(ego) \neq goal) \qquad (8)$$

Path constraint:
The path to parking lot follows the center line of the driving lane with max. deviation $dev_{max,st}$ in straight road segments and $dev_{max,cu}$ in curves.

$$\mathbf{G}((straight \rightarrow d \leq dev_{max,st}) \wedge (curve \rightarrow d \leq dev_{max,cu})) \qquad (9)$$

with: *straight, curve* boolean variables, true iff the road segment is straight (resp. curve); d is the distance from the center of the lane.

Control. The controller receives the reference path/trajectory from the path planner, and computes the steering and acceleration/deceleration commands, so that the vehicle moves along the path/trajectory. Vehicle should be kept within its dynamical limits with respect to its velocity, acceleration, jerk steering angle etc.
Vehicle moves along the reference path/trajectory:
The tracked path/trajectory shall not diverge from the reference path/trajectory more than d_{max}, e.g., 0.2 m, for any operating condition defined in the ODD.

$$\mathbf{G}\forall t(time(t) \wedge \mathsf{odd}(in)) \rightarrow (d_{\mathrm{error}}(t) \leq d_{\max}) \qquad (10)$$

where: $\mathsf{odd}(in)$ states that the condition in is in the ODD; $d_{\mathrm{error}}(t)$ is the maximal deviation between $path_{\mathrm{controlled}}(t)$ up to time t and the reference path $path_{\mathrm{ref}}$. The vehicle is within its dynamical limits:
The ego vehicle velocity v shall be bounded by v_{max}, and $v_{max,rev}$ for any operating condition in the ODD.

$$\mathbf{G}(\mathsf{odd}(in) \rightarrow (-v_{max,rev} \leq v \leq v_{max})) \qquad (11)$$

3.2   A Medical Autonomous System

In this section, we focus on the formalization of requirements, for an anaesthetic drug target control infusion system.

A patient's model is a component that predicts the future patient status of depth of anesthesia (site-effect concentration) based on the drug delivered. A model of the patient helps describing what has happened and what will happen with a planned dose for him/her. For intravenous drugs, the plasma concentration will be determined by the dose given (in weight units of drug), the distribution to different tissues in the body and the elimination from the body.

Let $x \in \mathbb{R}$ be the site effect concentration and

$$L := \{\text{Minimal-sedation}, \text{Sedation}, \text{Moderate-sedation}, \text{Deep-sedation}, \text{General-anaesthesia}\}$$

be the set of all possible levels in sedation in discretized form. The values "none", "less", "more" denote the amount of medicine to be used.

Formulas (12–14) prescribe the level of injection for the required sedation level:

$$\mathbf{G}\forall l \in L, (sedation\_req(l) \wedge x < low\_level(l)) \rightarrow \text{inject}(\text{more}) \tag{12}$$

$$\mathbf{G}\forall l \in L, (sedation\_req(l) \wedge x > upper\_level(l)) \rightarrow \text{inject}(\text{none}) \tag{13}$$

$$\mathbf{G}\forall l \in L, sedation\_req(l) \wedge low\_level(l) \leq x \leq upper\_level(l) \rightarrow \text{inject}(\text{less}) \tag{14}$$

where: $sedation\_req(l)$ is the required sedation level; $low\_level(l)$ and $upper\_level(l)$ are the lowest (resp. upper) level of drug for sedation level l.

The following formula describes how the site-effect concentration x diminishes over time when (via "inject(none)") not injecting medicine.

$$\mathbf{G}\forall t, t', ((time(t) \wedge concentration(x) \wedge$$
$$(\text{inject}(\text{none}) \; \mathcal{S} \; (time(t') \wedge concentration(x')))) \rightarrow$$
$$x = x' \times e^{(t'-t)/\tau}) \tag{15}$$

where $concentration(x)$ refers to the anaesthetic drug concentration at the current state and τ is a constant characterizing the speed of decay.

Equation (16) specifies how site-effect concentration x is raised over time for the required sedation level l, while further anaesthetic material is injected.

$$\mathbf{G}\forall l \in L, \; \forall t, t', \; \forall inj\_type, ((time(t) \wedge concentration(x) \wedge (inject(inj\_type) \wedge$$
$$(inj\_type = \text{less} \vee inj\_type = \text{more}) \; \mathcal{S} \; (time(t') \wedge concentration(x'))))$$
$$\rightarrow x = x' + (saturation\_level(l, inj\_type) - x') \times e^{(t'-t)/\tau}) \tag{16}$$

where $saturation\_level(l, inj\_type)$ is the desired saturation level of sedation l when injection is of type inj_type.

4 Related Work

In [28], the formalization of requirements for the runtime verification of an autonomous unmanned aircraft system was based on an extension of propositional LTL [29], where temporal operators are augmented with timing constraints. The Timed Quality Temporal Logic (TQTL) [4] has been proposed for expressing monitorable [27] spatio-temporal quality properties of perception systems based on NNs. TQTL is more limited in scope than our specification formalism, which has also the potential to express system-level requirements. The typed first-order logic [25] is an alternative, whose main difference from traditional first-order logic [26] is the explicit typing of variables, functions and predicates. This allows to reason about the domain of runtime control at the abstract level of types, instead of individual objects, thus reducing the complexity of monitoring.

5 Concluding Remarks

We presented a specification approach and a formalism for learning-enabled autonomous systems. To simplify the approach, yet make it general enough, we adopted the past *first-order* LTL, together with the first-order logic capabilities to use variables, functions, predicates and quantification over the domains of variables. We demonstrated the use of formalism on two safety-critical use cases.

Our approach abstracts away from the internals of the involved NNs, since the actual values of neurons, and the relations between them, are not part of the specification. Instead of these values, inspired by [4], we refer to the values and objects they represent. Research on how a NN actually maintains its ability to classify objects or perform other typical tasks is still undergoing, therefore abstracting away from it is a useful feature rather than a handicap.

A notable tradeoff in selecting the specification formalism exists (often related also to the model of execution) between the expressiveness and the ability to utilize it within different formal methods: testing, automatic (model-checking) and manual (theorem-proving) verification and monitoring (i.e., runtime verification). An important case of gaining decidability for scarifying the generality of the model and the formalism is that of the ability to perform automatic verification for propositional LTL (also for Computational Tree Logic CTL) of *finite state systems* [10,21]. Models for automatic verification hence often abstract the states as a Boolean combination. This helps achieving decidability and also taming down the complexity. Nevertheless, for actual systems, it is often desired to include data in both the specification and the model. For cyber-physical systems, and for learning-enabled autonomous systems in particular, the use of data and parametric specification are often essential. While comprehensive automatic verification needs then to be abandoned, it is still desired to apply testing and monitoring. These methods provide a weaker guarantee for correctness, but are still highly important in the system development process. The formalism that we proposed has the advantage to allow testing and runtime verification [7,17].

A common constraint on specification that we undertook is focusing on safety properties. Essentially, safety properties assert that "something bad never happens", whereas liveness properties assert, intuitively, that "something good will happen". A formal definition and proof that every property of linear execution sequences can be expressed as a conjunction of a safety and a liveness property appears in [1]. While temporal logic allows expressing safety and liveness properties, recently, there is a growing concentration on safety properties, abandoning liveness. One reason is that safety properties are *monitorable* in the sense that their violation can be detected within finite time. On the other hand, liveness properties may often be non-monitorable [8]: the fact that something "good" will happen can exist forever, not violated, yet that event or combination of states may be deferred forever. It turns out that automatic testability and monitorability for execution sequences with data exist for the kind of specification suggested here [7,17]. Furthermore, for cyber-physical systems, the requirement often involves setting some actual deadlines: the "good" thing to happen must occur within some given time (physical time or some virtual units of progress). Then, the property becomes a safety property: when that time is expired, a failure of the property happens, and by keeping up with the progress of time, one can monitor the failure after a finite amount of time.

References

1. Alpern, B., Schneider, F.B.: Recognizing safety and liveness. Distrib. Comput. **2**(3), 117–126 (1987)
2. Alshiekh, M., Bloem, R., Ehlers, R., Könighofer, B., Niekum, S., Topcu, U.: Safe reinforcement learning via shielding. In: AAAI 2018, pp. 2669–2678 (2018)
3. Apt, K.R., Kozen, D.: Limits for automatic verification of finite-state concurrent systems. Inf. Process. Lett. **22**(6), 307–309 (1986)
4. Balakrishnan, A., et al.: Specifying and evaluating quality metrics for vision-based perception systems. In: DATE, pp. 1433–1438 (2019)
5. Bartocci, E., Bloem, R., Maderbacher, B., Manjunath, N., Nickovic, D.: Adaptive testing for CPS with specification coverage. In: ADHS 2021 (2021)
6. Bartocci, E., Falcone, Y., Francalanza, A., Reger, G.: Introduction to runtime verification. In: Bartocci, E., Falcone, Y. (eds.) Lectures on Runtime Verification. LNCS, vol. 10457, pp. 1–33. Springer, Cham (2018). https://doi.org/10.1007/978-3-319-75632-5_1
7. Basin, D.A., Klaedtke, F., Müller, S., Zalinescu, E.: Monitoring metric first-order temporal properties. J. ACM **62**(2), 45 (2015)
8. Bauer, A., Leucker, M., Schallhart, C.: The good, the bad, and the ugly, but how ugly is ugly? In: Sokolsky, O., Taşıran, S. (eds.) RV 2007. LNCS, vol. 4839, pp. 126–138. Springer, Heidelberg (2007). https://doi.org/10.1007/978-3-540-77395-5_11
9. Bloem, R., et al.: RATSY – a new requirements analysis tool with synthesis. In: Touili, T., Cook, B., Jackson, P. (eds.) CAV 2010. LNCS, vol. 6174, pp. 425–429. Springer, Heidelberg (2010). https://doi.org/10.1007/978-3-642-14295-6_37
10. Clarke, E.M., Emerson, E.A.: Design and synthesis of synchronization skeletons using branching time temporal logic. In: Kozen, D. (ed.) Logic of Programs 1981.

LNCS, vol. 131, pp. 52–71. Springer, Heidelberg (1982). https://doi.org/10.1007/BFb0025774

11. Cordts, M., et al.: The cityscapes dataset for semantic urban scene understanding. CoRR, abs/1604.01685 (2016)

12. Donzé, A., Maler, O.: Robust satisfaction of temporal logic over real-valued signals. In: Chatterjee, K., Henzinger, T.A. (eds.) FORMATS 2010. LNCS, vol. 6246, pp. 92–106. Springer, Heidelberg (2010). https://doi.org/10.1007/978-3-642-15297-9_9

13. Fainekos, G.E., Pappas, G.J.: Robustness of temporal logic specifications. In: Havelund, K., Núñez, M., Roşu, G., Wolff, B. (eds.) FATES/RV 2006. LNCS, vol. 4262, pp. 178–192. Springer, Heidelberg (2006). https://doi.org/10.1007/11940197_12

14. Falcone, Y., Mounier, L., Fernandez, J.-C., Richier, J.-L.: Runtime enforcement monitors: composition, synthesis, and enforcement abilities. Formal Methods Syst. Des. 38(3), 223–262 (2011)

15. Ferrère, T., Nickovic, D., Donzé, A., Ito, H., Kapinski, J.: Interface-aware signal temporal logic. In: HSCC 2019, pp. 57–66 (2019)

16. Fowler, M., Distilled, U.M.L.: A Brief Guide to the Standard Object Modeling Language. Addison-Wesley, Boston (2004)

17. Havelund, K., Peled, D., Ulus, D.: First order temporal logic monitoring with BDDs. In: FMCAD 2017, pp. 116–123 (2017)

18. Hong, H.S., Lee, I., Sokolsky, O., Ural, H.: A temporal logic based theory of test coverage and generation. In: Katoen, J.-P., Stevens, P. (eds.) TACAS 2002. LNCS, vol. 2280, pp. 327–341. Springer, Heidelberg (2002). https://doi.org/10.1007/3-540-46002-0_23

19. Manna, Z., Pnueli, A.: Completing the temporal picture. Theor. Comput. Sci. 83, 91–130 (1991)

20. Nghiem, T., Sankaranarayanan, S., Fainekos, G., Ivancic, F., Gupta, A., Pappas, G.: Monte-Carlo techniques for falsification of temporal properties of non-linear hybrid systems. In: HSCC 2010, pp. 211–220 (2010)

21. Queille, J.P., Sifakis, J.: Specification and verification of concurrent systems in cesar. In: Dezani-Ciancaglini, M., Montanari, U. (eds.) Programming 1982. LNCS, vol. 137, pp. 337–351. Springer, Heidelberg (1982). https://doi.org/10.1007/3-540-11494-7_22

22. Prabhakar, P., Lal, R., Kapinski, J.: Automatic trace generation for signal temporal logic. In: RTSS 2018, pp. 208–217 (2018)

23. Redmon, J., Divvala, S., Girshick, R., Farhadi, A.: You only look once: unified, real-time object detection. In: CVPR 2016, pp. 779–788 (2016)

24. Roehm, H., Heinz, T., Mayer, E.C.: STLInspector: STL validation with guarantees. In: Majumdar, R., Kunčak, V. (eds.) CAV 2017, Part I. LNCS, vol. 10426, pp. 225–232. Springer, Cham (2017). https://doi.org/10.1007/978-3-319-63387-9_11

25. Beckert, B., Hähnle, R., Schmitt, P.H. (eds.): Verification of Object-Oriented Software. The KeY Approach. LNCS (LNAI), vol. 4334. Springer, Heidelberg (2007). https://doi.org/10.1007/978-3-540-69061-0

26. Smullyan, R.R.: First-Order Logic. Ergebnisse der Mathematik und ihrer Grenzgebiete. 2. Folge, Springer, Heidelberg (2012). https://doi.org/10.1007/978-3-642-86718-7

27. Balakrishnan, A., Deshmukh, J., Hoxha, B., Yamaguchi, T., Fainekos, G.: PerceMon: online monitoring for perception systems. In: Feng, L., Fisman, D. (eds.) RV 2021. LNCS, vol. 12974, pp. 297–308. Springer, Cham (2021). https://doi.org/10.1007/978-3-030-88494-9_18

28. Dutle, A., et al: From requirements to autonomous flight: an overview of the monitoring ICAROUS project. In: Proceedings of 2nd Workshop on Formal Methods for Autonomous Systems (FMAS). EPTCS, vol. 329, pp. 23–30 (2020)

29. Koymans, R.: Specifying real-time properties with metric temporal logic. Real-Time Syst.. **2**(4), 255–299 (1990)

NSV 2022

Verified Numerical Methods for Ordinary Differential Equations

Ariel E. Kellison[1]([✉])[iD] and Andrew W. Appel[2][iD]

[1] Cornell University, Ithaca, NY, USA
ak2485@cornell.edu
[2] Princeton University, Princeton, NJ, USA
appel@princeton.edu

Abstract. Ordinary differential equations (ODEs) are used to model the evolution of the state of a system over time. They are ubiquitous in the physical sciences and are often used in computational models with safety-critical applications. For critical computations, numerical solvers for ODEs that provide useful guarantees of their accuracy and correctness are required, but do not always exist in practice. In this work, we demonstrate how to use the Coq proof assistant to verify that a C program correctly and accurately finds the solution to an ODE initial value problem (IVP). Our verification framework is modular, and concisely disentangles the high-level mathematical properties expected of the system being modeled from the low-level behavior of a particular C program. Our approach relies on the construction of two simple functional models in Coq: a floating-point valued functional model for analyzing the intermediate-level behavior of the program, and a real-valued functional model for analyzing the high-level mathematical properties of the system being modeled by the IVP. Our final result is a proof that the floating-point solution returned by the C program is an accurate solution to the IVP, with a good quantitative bound. Our framework assumes only the operational semantics of C and of IEEE-754 floating point arithmetic.

1 Introduction

Computing accurate solutions to differential equations is a main topic in the field of numerical analysis. A typical problem in ordinary differential equations requires computing the numerical solution to autonomous initial value problems (IVPs) of the form

$$\frac{dx}{dt} = f(x), \quad x(t_0) = x_0 \tag{1}$$

at some time $t \in [t_0, T]$ to within a user-specified error tolerance. In this paper, our objective is to demonstrate a logical framework for verifying the accuracy and correctness of numerical programs that compute the solution to problems of the form (1). This framework is most suitable for critical applications that

O. Isac et al. (Eds.): NSV 2022/FoMLAS 2022, LNCS 13466, pp. 147–163, 2022.
https://doi.org/10.1007/978-3-031-21222-2_9

require guarantees of numerical accuracy (i.e., the numerical solution does not exceed the user-specified error tolerance) and program correctness (i.e., the particular implementation is bug-free and meets its specification). Our main result is a machine-checked theorem stating that a specific imperative implementation of a numerical method for the solution to an IVP produces a solution within a guaranteed error bound of the true solution, where the error bound accounts for two sources of error: discretization error and round-off error. We obtain this machine-checked theorem using a modular, layered approach to program verification that allows us to treat program correctness and each source of error separately within one logical framework, namely, the Coq proof assistant. For the results presented in this work, we have chosen the simple harmonic oscillator as an elementary but sufficiently illustrative example of an initial value problem[1]; for the numerical solution to this IVP, we consider a C implementation of the Störmer-Verlet ("leapfrog") method [1].

In contrast to *validated* numerical methods [2–4] and their implementations [5–7] which have a long history of deriving guaranteed error bounds for IVPs for ODEs, our framework for *verified* numerical methods for IVPs for ODEs has three distinct advantages for critical applications:

1. No additional computational overhead is introduced at run time.
2. Each source of error (e.g., discretization error, round-off error, data error, and bugs in the implementation) is treated separately in a modular way. This enables users to easily identify or emphasize areas of concern in their numerical method or program.
3. Guaranteed error bounds are directly connected to low-level properties of an implementation (in C or below). This connection provides assurance beyond the scope of validated methods.

To obtain a correctness-and-accuracy theorem that connects guaranteed error bounds to the low-level correctness of a C implementation of the leapfrog method, we *layer* the verification using several tools and libraries that are fully integrated into the Coq proof assistant. In particular, we prove that the C program refines a functional model using VST [8], and (separately) prove that the functional model has the desired properties using the Coquelicot formalization of real analysis [9], the Coq Interval [10] package, and VCFloat [11,12]. Our Coq development is available at github.com/VeriNum/VerifiedLeapfrog.

2   Main Result

Our main objective is to verify that a C implementation of leapfrog integration (given in Fig. 1) is correct, and that it accurately solves the system of ordinary differential equations for the simple harmonic oscillator in $\mathbb{R}^2$ to within an accuracy acc at time T. In particular, we consider the system of equations

---

[1] This particular model problem admits an analytical solution and is therefore not expected to be of practical interest on its own. Instead, it is chosen for demonstrating and analyzing the performance of our logical framework.

$$\frac{dp}{dt} = -\omega^2 q, \quad \frac{dq}{dt} = p, \tag{2}$$

where ω, p, and q are, respectively, the frequency, momentum, and position of the oscillator. To indicate that two functions $p : \mathbb{R} \to \mathbb{R}$ and $q : \mathbb{R} \to \mathbb{R}$ with initial conditions $p(t_0) = p_0$ and $q(t_0) = q_0$ constitute the continuous system (2) we use the predicate[2] Harmonic_oscillator_system ω p q,

Definition Harmonic_oscillator_system $(\omega : \mathbb{R})$ $(p\ q : \mathbb{R} \to \mathbb{R})$: Prop := smooth_fun $p \wedge$ smooth_fun $q \wedge \forall t : \mathbb{R}$, (Derive_n q 1 $t = p\ t \wedge$ Derive_n p 1 $t = F(q(t), \omega))$.

where the predicate (smooth_fun f) indicates that f is continuously differentiable and (Derive_n f n x) is the Coquelicot abstraction for the nth derivative of f at x; the function $F(q(t), \omega)$ is the restoring force acting on the system:

Definition F $(x\ \omega : \mathbb{R}) : \mathbb{R} := -\omega^2 \cdot x$.

An integer-step leapfrog discretization of the continuous system (2) on a time interval $[0, T]$ uniformly partitioned by a fixed time step h with unit frequency $\omega = 1$ updates the position q and momentum p of the oscillator as

$$q_{n+1} = q_n + h p_n - \frac{h^2}{2} q_n \tag{3}$$

$$p_{n+1} = p_n - \frac{h}{2}(q_n + q_{n+1}). \tag{4}$$

```
struct state {float p, q;};

float force(float q) { return -q; }

void integrate(struct state *s) {
    int n, N=1000; float t, h = 1.0f / 32.0f;
    s->q = 1.0f; s->p = 0.0f; t = 0.0f;
    for (n = 0; n < N; n++) {
        float a = force(s->q);
        s->q = s->q + h * s->p + (0.5f * (h * h)) * a;
        s->p = s->p + (0.5f * h) * (a + force(s->q));
        t = t + h;
} }
```

Fig. 1. Leapfrog integration of the harmonic oscillator implemented in C with time step $h = \frac{1}{32}$, frequency $\omega = 1$, initial conditions $(p_0, q_0) = (0, 1)$.

If we define the *global error* at the nth time step $t_n = nh \le T$ of leapfrog integration as the residual between the ideal solution $(p(t_n), q(t_n))$ and the numerical

---

[2] The form **Definition** *name* (*arguments*) : *type* := *term* in Coq binds *name* to the value of the *term* of type *type*; Prop is the type of well-formed propositions.

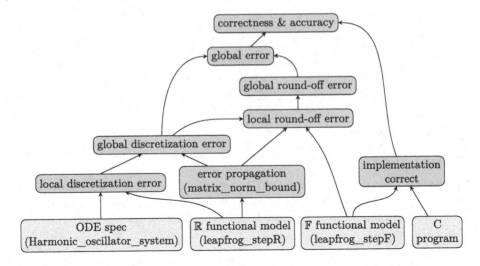

Fig. 2. Theorem dependency.

solution (p_n, q_n), i.e., $E_n = \|(p(t_n), q(t_n)) - (p_n, q_n)\|$, then the C implementation of Eqs. (3–4) is accurate if it has global error $E_n \leq$ acc (Fig. 2).

We prove the accuracy and correctness of the C implementation by composing several proofs: that the C program correctly implements a floating-point functional model; that in each iteration the floating-point functional model accurately approximates a real-valued functional model; that in each iteration the real-valued model accurately approximates the continuous ODE; that per-iteration errors are uniformly bounded by a propagation factor; and that the global propagation of per-iteration errors is bounded above by the desired accuracy. The main theorem then proves, from the composition of all of these theorems, and assuming only the operational semantics of C and of IEEE-754 floating point arithmetic, that the floating-point solution returned by the C program shown in Fig. 1 is an accurate solution to the ODE, with a good quantitative bound.

We encapsulate the expected floating-point behavior of the C function integrate of Fig. 1 on input $(p, q) = ic \in \mathbb{F}^2$ using the a floating-point valued functional model (leapfrog_stepF h ic), given in Fig. 3. We reason about the behavior of leapfrog integration in exact arithmetic by defining a real-valued functional model (leapfrog_stepR h ic). Iterations of leapfrog_stepF and leapfrog_stepR are defined as iternF and iternR. Henceforth we assume $\omega = 1$ and we omit it.

We use the predicate (accurate_harmonic_oscillator acc x n) to indicate that the single-precision floating-point valued momentum-position pair x differs by at most acc from the true position and momentum (at time $T = Nh$) of the ideal system defined by Eq. 2. Then, with x being the result computed by the C program, the C program specification is stated as integrate_spec:

Definition integrate_spec :=
  DECLARE _integrate
  WITH s: val

Definition leapfrog_stepR
$(h : \mathbb{R})$ $(ic : \mathbb{R}^2)$: $\mathbb{R}^2 :=$
  let p := fst ic in let q := snd ic in
  let $q' := (1 - \frac{h^2}{2}) \cdot q + h \cdot p$ in
  let $p' := (1 - \frac{h^2}{2}) \cdot p - \frac{h}{2} \cdot (2 - \frac{h^2}{2}) \cdot q$ in
  (p', q').

Fixpoint iternR
$(h : \mathbb{R})$ $(ic : \mathbb{R}^2)$ $(n: \mathbb{N})$: $\mathbb{R}^2 :=$
  match n **with**
  | 0 $\Rightarrow$ ic
  | S $n' \Rightarrow$
     iternR h (leapfrog_stepR h ic) n'
end.

Definition F (x : $\mathbb{F}$) : $\mathbb{F} := -x$.
Definition leapfrog_stepF
$(h: \mathbb{F})$ $(ic : \mathbb{F}^2)$: $\mathbb{F}^2 :=$
  let p := fst ic in let q := snd ic in
  let $q' := (q + h \cdot p) + (\frac{1}{2} \cdot (h \cdot h)) \cdot F(q)$ in
  let $p' := p + (\frac{1}{2} \cdot h) \cdot (F(q) + F(q'))$ in
  (p', q').

Fixpoint iternF
$(h: \mathbb{F})$ $(ic: \mathbb{F}^2)$ $(n: \mathbb{N})$: $\mathbb{F}^2 :=$
  match n **with**
  | 0 $\Rightarrow$ ic
  | S $n' \Rightarrow$
     iternF h (leapfrog_stepF h ic) n'
end.

Fig. 3. The floating-point and real valued functional models for leapfrog integration of the harmonic oscillator.

PRE [tptr t_state] PROP() PARAMS(s) SEP(data_at_ Tsh t_state s)
POST [tvoid]   EX ($x: \mathbb{F} \times \mathbb{F}$), PROP(accurate_harmonic_oscillator acc x N)
                  RETURN() SEP(data_at Tsh t_state x s).

The precondition and postcondition are assertions about any C value s that is the address of a struct state. In particular,

PRE: The precondition asserts that the function parameter (of type pointer-to-struct-state) does indeed contain the value s and that the "data at" that location is uninitialized (or is initialized but we don't care).

POST: The postcondition asserts that a pair x of single-precision floating-point values that are an accurate solution to the ODE are stored at address s.

If the C function satisfies this specification, then it correctly implements an accurate numerical integration of the ODE, which is our desired main result. We denote the C function's abstract-syntax tree as f_integrate and prove the main theorem body_integrate, which guarantees that f_integrate satisfies the specification integrate_spec:

Theorem body_integrate : semax_body Vprog Gprog f_integrate integrate_spec.

In the remainder of the paper, we present the modular proofs of accuracy and correctness that are composed to derive this main result.

3   Verified Error Bounds

For a given accuracy acc, time step h, initial condition (p_0, q_0), and final time $T = Nh$, our goal is to prove that the solution $(\hat{p}_N, \hat{q}_N)$ obtained by the C implementation of Eqs. (3–4) given in Fig. 1, has global error E_N bounded above by acc:

$$E_N = \|(p(t_N), q(t_N)) - (\hat{p}_N, \hat{q}_N)\| \leq \text{acc}. \tag{5}$$

We derive a verified upper bound for E_N by considering separately the global discretization error and global round-off error. If we denote the numerical solution in ideal arithmetic at time t_N as $(\tilde{p}_N, \tilde{q}_N)$, then an upper bound on the global error is

$$E_N = \|(p(t_n), q(t_N)) - (\hat{p}_N, \hat{q}_N)\| \tag{6}$$
$$\leq \underbrace{\|(p(t_N), q(t_N)) - (\tilde{p}_N, \tilde{q}_N)\|}_{\text{global discretization error}} + \underbrace{\|(\tilde{p}_N, \tilde{q}_N) - (\hat{p}_N, \hat{q}_N)\|}_{\text{global round-off error}} = D_N + R_N.$$

We obtain bounds on the global discretization error D_N and global round-off error R_N by first estimating the maximum possible *local error* from each source and then estimating the propagation of local errors as the iterations advance.

The local error associated with a numerical method is the residual between the ideal solution and the numerical solution after a single time step of size h starting from the same initial point [13]. To estimate the local discretization error τ_d at time $t_n = nh$ we therefore analyze the residual $\|(p(t_n), q(t_n)) - (\tilde{p}_n, \tilde{q}_n)\|$ where p and q satisfy (Harmonic_oscillator_system ω p q) and $(\tilde{p}_n, \tilde{q}_n)$ is defined as (leapfrog_stepR h $(p(t_{n-1}), q(t_{n-1}))$). Similarly, we estimate the local round-off error τ_r by analyzing the residual $\|(\tilde{p}_n, \tilde{q}_n) - (\hat{p}_n, \hat{q}_n)\|$ where $(\tilde{p}_n, \tilde{q}_n)$ is defined as (leapfrog_stepR h $(\hat{p}_{n-1}, \hat{q}_{n-1})$) and $(\hat{p}_n, \hat{q}_n)$ is defined as the injection of (leapfrog_stepF h $(\hat{p}_{n-1}, \hat{q}_{n-1})$) into the reals.

Deriving bounds on the global errors R_N and D_N requires that we are able to invoke our local error theorems at any iteration $0 \leq n \leq N$. We therefore conservatively estimate the local errors τ_d and τ_r such that

$$\max_{n \in [N]} \|(p(t_n), q(t_n)) - (\tilde{p}_n, \tilde{q}_n)\| \leq \tau_\text{d}, \quad \text{and} \tag{7}$$

$$\max_{n \in [N]} \|(\tilde{p}_n, \tilde{q}_n) - (\hat{p}_n, \hat{q}_n)\| \leq \tau_\text{r}. \tag{8}$$

We derive such a τ_d and τ_r using the fact that the momentum p and position q of both the ideal solution specified by Harmonic_oscillator_system and the numerical solution specified by leapfrog_stepR do not grow too large on the finite time interval $t_0 \leq t_n \leq T$. In particular, observe that one can prove from the specification Harmonic_oscillator_system (even without solving the ODE) that $\|(p(t), q(t))\| = \|(p_0, q_0)\|$ for all t. Unfortunately, this property does not hold *exactly* for leapfrog_stepR but we prove bounds on the growth of $\|$leapfrog_stepR h p q $\|$ for all p and q—see Sect. 3.2. Finally, while the exact conservation of $\|(p(t), q(t))\|$ in our model problem is useful for deriving tight bounds on local errors, it is not a requirement of our analysis. The error analysis presented here applies to IVPs of the form (1) as long as the local errors can be uniformly bounded on the finite time interval of concern, which is guranteed provided that $f(x)$ is Lipschitz continuous in x [14].

3.1   Local Discretization Error

Local discretization error is estimated as the residual difference between the exact solution characterized by Harmonic_oscillator_system and the numerical solution computed in exact arithmetic by (leapfrog_stepR) starting from the point $(p(t_n), q(t_n)) \in \mathbb{R}^2$ after a single time step of size h. We prove that the local discretization error is bounded, for all t, by $\tau_{\mathrm{d}} = h^3 \, \|(p(t_0), q(t_0))\|$.

Theorem local_discretization_error :
$\quad \forall \, (p \, q : \mathbb{R} \to \mathbb{R}) \, (t_0 \, t_n \, h : \mathbb{R}), \, 0 < h \leq 4 \to$
$\quad$ **let** $\omega := 1$ **in** Harmonic_oscillator_system $\omega \, p \, q \to$
$\quad$ **let** $(p_n, q_n) :=$ leapfrog_stepR $h \, (p(t_n), q(t_n))$ **in**
$\quad \|(p(t_n + h), q(t_n + h)) - (p_n, q_n)\| \leq h^3 \, \|(p(t_0), q(t_0))\|.$

Proof. We expand the ideal solution of the harmonic oscillator $(p(t_n + h),$ $q(t_n + h))$ as Taylor expansions around t_n using the Taylor_Lagrange theorem from the Coquelicot library [9] and use the derivative relations for p and q from Harmonic_oscillator_system to derive the differences

$$|p(t_n + h) - p_n| = h^3 \left| \frac{p(\eta_1)}{4!} - \frac{q(t_n)}{12} \right|, \tag{9a}$$

$$|q(t_n + h) - q_n| = \frac{h^3}{3!} |p(\eta_2)| \tag{9b}$$

for some $t_n < \eta_1, \eta_2 < t_n + h$. Recall that $\|(p(t), q(t))\| = \|(p(t_0), q(t_0))\|$ is a property of our model problem. Provided that $0 < h \leq 4$, it then follows that

$$\|(p(t_n), q(t_n)) - (p_n, q_n)\| \leq \tau_{\mathrm{d}} = h^3 \, \|(p_0, q_0)\|. \tag{10}$$

We will see in the next section that the restriction $h \leq 4$ is not overly restrictive.

3.2   Propagation of Errors

To bound the propagation of local errors over n iterations, we use the 2-norm of the transition matrix of leapfrog updates to position and momentum [15–17]. In particular, if we represent the leapfrog method for the evolution of the harmonic oscillator as the transition matrix $M(h) : (p_n, q_n) \mapsto (p_{n+1}, q_{n+1})$

$$M(h) = \begin{pmatrix} 1 + \zeta & \frac{\zeta}{h} (2 + \zeta) \\ h & 1 + \zeta \end{pmatrix} \quad \text{with} \quad \zeta = -\frac{h^2}{2}, \tag{11}$$

then the evolution over n steps is denoted as applications of powers of the transition matrix $M(h)$ to the initial conditions p_0 and q_0:

$$\begin{pmatrix} p_n \\ q_n \end{pmatrix} = (M(h))^n \begin{pmatrix} p_0 \\ q_0 \end{pmatrix}. \tag{12}$$

An upper bound for the global error e_n (where e_n could be either R_n or D_n) at step n can be decomposed into two parts: the local error at step n and the propagation of accumulated errors from previous steps.

$$e_n = \|(p(t_n), q(t_n)) - (p_n, q_n)\| = \|(p(t_n), q(t_n)) - M(h)(p_{n-1}, q_{n-1})\| \qquad (13)$$

$$\leq \underbrace{\|(p(t_n), q(t_n)) - M(h)(p(t_{n-1}), q(t_{n-1}))\|}_{\text{local error at step } n} +$$

$$\underbrace{\|M(h)(p(t_{n-1}), q(t_{n-1})) - M(h)(p_{n-1}, q_{n-1})\|}_{\text{propagation of prior local errors}}$$

$$\leq \tau + \|M(h)\| \, \|(p(t_{n-1}), q(t_{n-1})) - (p_{n-1}, q_{n-1})\| \, .$$

We can therefore estimate an upper bound for e_n from an appropriate estimate for the local error at step n and a reasonable approximation of $\|M(h)\|$.

To use Eq. (13) in our verified error analysis, we define $M(h)$ using the Coquelicot matrix library. We then derive a tight bound on $\|M(h)\|_2$ using the predicate two_norm_pred to indicate that the real number σ is the 2-norm of the $n \times n$ matrix A:

Definition two_norm_pred $(n: \mathbb{N})$ $(A : \text{matrix } \mathbb{C} \; n \; n)$ $(\sigma: \mathbb{R})$: Prop :=
$\forall \, (u : \text{vector } \mathbb{C} \; n \,)$, $\|Au\| \leq \sigma \|u\| \wedge (\neg \exists \, (s : \mathbb{R}), \forall$
$(x : \text{vector } \mathbb{C} \; n)$, $\|Ax\| \leq s \|x\| < \sigma \|x\|)$.

We prove (but do not present here) that this predicate is satisfied for any matrix $A \in \mathbb{R}^{2 \times 2}$ by the maximum singular value of A. $\|M(h)\|$ is therefore defined as the positive real number $\sigma(h)$ in (two_norm_pred 2 $(M(h))$ $(\sigma(h))$) such that $\sigma(h)$ is the square root of the maximum eigenvalue of the matrix $B = \overline{M}(h)^T M(h)$:

Definition σ $(h : \mathbb{R})$: $\mathbb{R}$:=
  let $a := \sqrt{h^6 + 64}$ **in**
  let $A := (h^{10} + h^7 a + 4h^6 + 64h^4 + 4h^3 a + 32ha + 256)(h^2 - 2)^2$ **in**
  $\sqrt{A / (2(h^4 - 4h^2 + 4)((-h^3 + 8h + a)^2 + 16(h^2 - 2)^2))}$

Leapfrog integration of the harmonic oscillator with unit frequency is *stable* for time-steps $0 < h < \sqrt{2}$ [18]; since our time step is fixed by $h = \frac{1}{32}$ in our C program, we derive a verified bound on the solution vector (p_n, q_n) at step n for any initial (p_0, q_0) by proving the following theorem, which follows by induction on the iteration number and unfolding of the definition of the predicate for the 2-norm.

Theorem matrix_bound : $\forall$ $(p_0 \; q_0: \mathbb{R})$ $(n : \mathbb{N})$, $\|(M(h))^n(p_0, q_0)\| \leq (\sigma(h))^n \|(p_0, q_0)\|$.

A bound on leapfrog_stepR follows as a corollary. In particular, for $h = \frac{1}{32}$, we have $\sigma(h) \leq 1.000003814704543$, and therefore

Corollary method_norm_bound :
  $\forall \, p \; q: \mathbb{R}$, $\|(\text{leapfrog\_stepR}(p, q)h)\| \leq 1.000003814704543 \, \|(p, q)\|$.

Given that the C program (Fig. 1) runs for $N = 1000$ iterations with the initial condition $(p(t_0), q(t_0)) = (0, 1)$, method_norm_bound guarantees that each component of the numerical solution (position and momentum) will be bounded by 1.00383 in absolute value; we use these bounds on the solution vector when deriving an upper bound on the local round-off error.

3.3 Global Discretization Error

Analyzing the recurrence in the term for the propagation of prior local errors in Eq. (13) over several iterations leads to the following estimate of an upper bound for the global discretization error.

$$D_N = \|(p(t_n), q(t_n)) - (\text{iternR h } (p(t_0), q(t_0)) \text{ n})\| \le h^3 \|(p(t_0), q(t_0))\| \sum_{k=0}^{n-1} \sigma(h)^k.$$

We prove that this estimate holds by invoking our local error theorem and performing induction on the iteration step in the following theorem.

Theorem global_discretization_error :
  $\forall (p \; q : \mathbb{R} \to \mathbb{R}) \; (t_0 : \mathbb{R})$, **let** $\omega := 1$ **in**
  Harmonic_oscillator_system $\omega \; p \; q \to$
  $\forall \; n : \mathbb{N}$,   **let** $t_n := t_0 + nh$ **in**

$$\|(p(t_n), q(t_n)) - (\text{iternR } h \; (p(t_0), q(t_0)) \text{ n})\| \le h^3 \|(p(t_0), q(t_0))\| \sum_{k=0}^{n-1} \sigma(h)^k.$$

Given that the C program (Fig. 1) runs for $N = 1000$ iterations with time step $h = \frac{1}{32}$ and initial condition $(p(t_0), q(t_0)) = (0, 1)$, the contribution from discretization error to the global error at $t = Nh$ is guaranteed to be at most $3.06 \cdot 10^{-2}$.

3.4 Local Round-Off Error

Local round-off error is the residual difference between the numerical solution computed in exact arithmetic and the numerical solution computed in single-precision floating-point arithmetic after a single time step of size $h = \frac{1}{32}$ on the same input. We derive a bound on the maximum possible local round-off error for leapfrog integration of the harmonic oscillator using VCFloat [11, 12] and the Coq interval package [10]:

Theorem local_roundoff_error:
  $\forall \; x : $ state, boundsmap_denote leapfrog_bmap (leapfrog_vmap x) $\to$
  $\|$FT2R_prod(leapfrog_stepF $h \; x$) $-$ leapfrog_stepR h (FT2R_prod x)$\| \le \tau_r$,
  where $\tau_r = 1.399 \cdot 10^{-7}$.

The proof of local_roundoff_error is mostly automatic. We will not show the details here; see Ramanandro et al. [12] and Appel and Kellison [11]. The function FT2R_prod : $\mathbb{F} \times \mathbb{F} \to \mathbb{R} \times \mathbb{R}$ in local_roundoff_error injects floating-point pairs to real number pairs. The boundsmap_denote hypothesis enforces bounds on the components of the state vector $x = (p, q) \in \mathbb{F}^2$. In particular, we have constructed leapfrog_bmap to specify that $-1.0041 \le p, q \le 1.0041$.

Tighter bounds on p and q will result in a tighter round-off error bound. Initially, $\|(p_0, q_0)\| = 1$, so $-1 \leq p_0, q_0 \leq 1$. But as errors (from discretization and round-off) accumulate, the bounds on p, q must loosen.

In principle the leapfrog_bmap could be a function of n; the nth-iteration bounds on p, q could be used to calculate the nth-iteration round-off error. As discussed at the beginning of Sect. 3, we prove a local round-off error bound τ just once, in part because the VCFloat library requires the bounds to be constant values. So τ is basically the worst-case bound on p, q after the last iteration.

The looser the bounds, the worse the round-off error, therefore the looser the bounds. Fortunately the round-off error is only weakly dependent on the bounds on p and q, so we can cut this Gordian knot by choosing τ_r small enough to prove an adequately tight bound in local_roundoff_error and large enough to be proved from global_roundoff_error.

We derive the hypothesis $-1.0041 \leq p, q \leq 1.0041$ as follows. If there were no round-off error, then (according to theorem method_norm_bound), $\|(p, q)\|$ increases by (at most) a factor of 1.0000039 in each iteration, so over 1000 iterations that is (at most) 1.00383. The machine epsilon for single-precision floating point is $\epsilon = 1.19 \cdot 10^{-7}$. Assuming this error in (each component of) the calculation of $\|(p, q)\|$, then the norm of the floating-point solution for N iterations can be bounded as:

$$\|(\text{leapfrog\_stepF } x)\| \leq \|(\text{leapfrog\_stepR } x\| + \epsilon \sum_{k=0}^{N-1} \sigma(h)^k$$

$$\leq \sigma(h)^N + \epsilon \sum_{k=0}^{N-1} \sigma(h)^k \leq 1.0041. \qquad (14)$$

Finally, note that the appropriate application of the function FT2R_prod has been elided in Eq. (14) for succinctness—e.g., $\|\text{leapfrog\_stepR}(x)\|$ should appear as $\|\text{FT2R\_prod}(\text{leapfrog\_stepR}(x))\|$—we will continue to omit this function in the remainder of the paper.

3.5   Global Round-Off Error

We estimate an upper bound for the global round-off error R_N by replicating the analysis for the global discretization error D_N given in Sect. 3.3.

Theorem global_roundoff_error :
   boundsmap_denote leapfrog_bmap (leapfrog_vmap (p_0, q_0)) $\rightarrow$
   $\forall\, (n: \mathbb{N}),\ n \leq N \rightarrow$
   boundsmap_denote leapfrog_bmap (leapfrog_vmap (iternF h (p_0, q_0) n))
   $\wedge \|$ (iternR h (p_0, q_0) n) $-$ (iternF h (p_0, q_0) n) $\| \leq \tau_r \sum_{k=0}^{n-1} \sigma(h)^k.$

Theorem global_roundoff_error provides a verified bound for global round-off error. It states that if the bounds required by the boundsmap_denote predicate (see Sect. 3.4) hold, then the solution (iternF h (p_0, q_0) n) obtained by single precision leapfrog integration over N iterations satisfies the bounds required by boundsmap_denote, and that the global round-off error for N iterations is upper bounded by the product of the maximum local round-off error and the sum of powers of the global error propagation factor (see Sect. 3.2).

The upper bound on $\|$ (iternR h (p_0, q_0) $n)$ $-$ (iternF h (p_0, q_0) $n)$ $\|$ follows by induction: if we define the floating-point solution after n steps of integration as $(\hat{p}, \hat{q}) = $ (iternF h (p_0, q_0) n) then

$$\|\text{iternR } h \ (p_0, q_0) \ (n+1) - \text{iternF } h \ (p_0, q_0) \ (n+1)\|$$
$$= \|\text{iternR } h \ (p_0, q_0) \ (n+1) - \text{leapfrog\_stepF } (\hat{p}, \hat{q})\|$$
$$\leq \underbrace{\|\text{iternR } h \ (p_0, q_0) \ (n+1) - \text{leapfrog\_stepR } (\hat{p}, \hat{q})\|}_{\text{propagation of prior local errors}} +$$
$$\underbrace{\|(\text{leapfrog\_stepR } (\hat{p}, \hat{q}) - \text{leapfrog\_stepF } (\hat{p}, \hat{q})\|}_{\text{local round-off error at step } (n+1)}$$
$$\leq \tau_{\text{r}} \sum_{k=0}^{n} \sigma(h)^k.$$

$$(15)$$

From Eq. 15 it is clear that we must invoke local_roundoff_error in the proof of global_roundoff_error. To do so, we must show that $(\hat{p}, \hat{q})$ satisfy the boundsmap_denote predicate. To this end, we prove lemma itern_implies_bmd, which guarantees that the estimate in Eq. (14) is sufficient.

Lemma itern_implies_bmd:
$\forall \ (p \ q \colon \mathbb{F}) \ (n \colon \mathbb{N}), \ n+1 \leq N \to$
boundsmap_denote leapfrog_bmap (leapfrog_vmap (iternF h (p,q) n)) $\to$
$\| $ (iternR h (p,q) $(n+1)$) $-$ (iternF h (p,q) $(n+1)$) $\| \leq \tau_{\text{r}} \sum_{k=0}^{n} \sigma(h)^k \to$
$\| $ (iternR h (p,q) $(n+1)$)$\| \leq \sigma(h)^N \to$
boundsmap_denote leapfrog_bmap (leapfrog_vmap (iternF h (p,q) $(n+1)$)).

From global_roundoff_error we conclude that the contribution from round-off error to the global error at $t = Nh$ is guaranteed to be at most $1.4 \cdot 10^{-4}$.

3.6   Total Global Error

Using global_roundoff_error and global_discretization_error   from Sects. 3.1 and 3.4, we derive a verifed concrete upper bound for the total global error for single precision leapfrog integration of the harmonic oscillator over N time steps as

$$E_N \leq \underbrace{\|(p(t_N), q(t_N)) - (\tilde{p}_N, \tilde{q}_N)\|}_{\text{global discretization error}} + \underbrace{\|(\tilde{p}_N, \tilde{q}_N) - (\hat{p}_N, \hat{q}_N)\|}_{\text{global round-off error}}$$
$$\leq (\tau_{\text{d}} + \tau_{\text{r}}) \sum_{k=0}^{N-1} \sigma(h)^k \leq 0.0308. \quad (16)$$

This bound is guaranteed by the following theorem, which uses the closed form expression for the geometric series in Eq. 16.

Theorem total_error:
$\forall\,(p_t\ q_t\colon \mathbb{R}\to\mathbb{R})\,(n:\mathbb{N}),\, n \le N \to$
let $t_0 := 0$ **in let** $t_n := t_0 + nh$ **in** $p_t(t_0) = p_0\;\;\to\;\;q_t(t_0) = q_0 \to$
let $\omega := 1$ **in** Harmonic_oscillator_system $\omega\ p_t\ q_t \to$
$\|(p_t(t_n), q_t(t_n)) - (\text{iternF } h\ (p_0, q_0)\ n)\| \;\le\; (\tau_\mathrm{d} + \tau_\mathrm{r})(\sigma(h)^n - 1)/(\sigma(h) - 1)$.

In the following sections, we describe how the bound provided by total_error is composed with the refinement proof that C program implements the floating-point functional model to prove our main result.

4   Program Verification

The Verified Software Toolchain [19] is a program logic for C, with a soundness proof in Coq with respect to the formal operational semantics of C, and with proof automation tools in Coq for interactive verification.

When verifying programs in VST (or in other program logics), it is common to *layer* the verification: prove that the C program refines a functional model, and (separately) prove that the functional model has the desired properties. In this case, the functional model is our floating-point model defined by the functions leapfrog_stepF and iternF .

We showed a *high-level* specification for the integrate function of the C program (Fig. 1) in Sect. 2; namely, that it accurately solves the ODE. Here we start with the *low-level* spec that the C program implements the floating-point functional model:

Definition integrate_spec_lowlevel :=
  DECLARE _integrate
  WITH s: val
  PRE [tptr t_state]
     PROP(iternF_is_finite) PARAMS (s) SEP(data_at_ Tsh t_state s)
  POST [tvoid]
     PROP() RETURN()
     SEP(data_at Tsh t_state (floats_to_vals (iternF h (p_init,q_init) N)) s).

This claims that when the function returns, the float values iternF h $(p_{\text{init}}, q_{\text{init}})$ N will be stored at location s—provided that (in the precondition) struct-fields s->p and s->q are accessible, and assuming iterF_is_finite .

The functional model is deliberately designed so that its floating-point operations adhere closely to the operations performed by the C program. So the proof is almost fully automatic, except that the VST user must provide a loop invariant. In this case the loop invariant looks much like the function postcondition, except with the iteration variable n instead of the final value N.

The proofs of these functions are fairly short; see Table 1. If the program had used nontrivial data structures or shared-memory threads, the functional model might be the same but the C program would be more complex. The relation between the program and the model would be more intricate. VST can handle such intricacy with more user effort.

Table 1. VST proof effort, counting nonblank, noncomment lines of C or Coq. Proofs count text between (not including) **Proof** and **Qed**

C program	C lines	Proof lines	Proof chars
force	3	2	25
lfstep	6	15	281
integrate	12	30	953

We designed the functional model so that one can prove correctness of the C program without knowing (almost) anything about the properties of floating-point, and (completely) without knowing about the existence of the real numbers. One is simply proving that the C program does these float-ops, in this tree-order, without needing to know why.

5   Composing the Main Theorems

We prove subsume_integrate: that integrate_spec_lowlevel implies the high-level integrate_spec. We use the theorem yes_iternF_is_finite to discharge the precondition iternF_is_finite of the low-level spec, and use the total_error theorem to show that the iternF postcondition of the low-level spec implies the accurate_harmonic_oscillator postcondition of the high-level spec.

Lemma subsume_integrate:
    funspec_sub (**snd** integrate_spec_lowlevel) (**snd** integrate_spec).

The proof is only a few lines long (since all the hard work is done elsewhere). Then, using VST's subsumption principle [20] we can prove the body_integrate theorem stated in Sect. 2.

6   Soundness

Underlying our main result are several soundness theorems: soundness of VST [19] with respect to the formal operational semantics of C and the Flocq [10] model of IEEE-754 floating point; soundness of the Interval package and the VCFloat package with respect to models of floating point and the real numbers; proofs of Coquelicot's standard theorems of real analysis.

To test this, we used Coq's **Print Assumptions** command to list the axioms on which our proof depends. We use 6 standard axioms of classical logic (excluded middle, dependent functional extensionality, propositional extensionality) and 74 axioms about primitive floats and primitive 63-bit integers.

Regarding those 74: One could compute in Coq on the binary model of floating-point numbers, with no axioms at all. Our proofs use such reasoning. However, the standard installation of the Interval package allows a configuration that uses the Coq kernel's support for native 64-bit floating-point and 63-bit modular integers—so they appear in our list of trusted axioms whether we use them or not. The algorithms within Interval and VCFloat (written as functional programs in Coq's Gallina language) would run much faster with machine floats and machine integers. But then we would have to trust that Coq's kernel uses them correctly, as claimed by those 74 axioms.

7 Related Work

A significant difference between the logical verification framework presented in this paper and the majority of existing methods for estimating the error in numerical solvers for differential equations is that our verification framework connects guaranteed error bounds to low-level properties of the solver implementation. An exception is the work by Boldo *et al.* [21], which verifies a C program implementing a second-order finite difference scheme for solving the one-dimensional acoustic wave equation. Although their model problem is a PDE, the framework could be generalized to IVPs for ODEs. The authors derive a total error theorem that composes global round-off and discretization error bounds, and connect this theorem to a proof of correctness of their C program. The authors use a combination of tools to perform their verification, including Coq, Gappa, Frama-C, Why, and various SMT solvers. Unlike VST, Frama-C has no soundness or correctness proof with respect to any formal semantics of C. Furthermore, VST is embedded in Coq and therefore enjoys the expressiveness of Coq's high-order logic; Frama-C lacks this expressivity, and this point was noted by the authors as a challenge in the verification effort.

The leapfrog method used as a solver for the two-dimensional model IVP in this paper is a simple example of one of many different families of solvers for IVPs for ODEs. Another class of methods that have been studied using logical frameworks and their related tools are Runge-Kutta methods. Boldo *et al.* [22] analyze the round-off errors (but not discretization error) of Runge-Kutta methods applied to linear one-dimensional IVPs using Gappa [23,24], a tool for bounding round-off error in numerical programs that produces proof terms which can be verified in Coq. The authors use Gappa to derive tight local error bounds, similar to our use of VCFloat as described in Sect. 3.4, but perform their global round-off error analysis outside of a mechanized proof framework. Immler and Hölzl [25] formalize IVPs for ODEs in Isabelle/HOL and prove the existence of unique solutions. They perform an error analysis on the simplest one-dimensional Runge-Kutta method and treat discretization error and round-off error uniformly as perturbations of the same order of magnitude.

Finally, as previously mentioned, validated numerical methods for ODEs have a long history of using interval arithmetic to derive guaranteed estimates for global truncation and round-off error [2,4,26–29]; this can be computationally inefficient for practical use. However, even unvalidated methods for estimating global error are inefficient. A common approach entails computing the solution a second time using a smaller time step, and using this second computation as an approximation of the exact solution [30]. An alternative approach implements *a posteriori* global error estimates in existing ODE solvers [31,32] to control errors by dynamically adjusting the time-step. Unlike the error bounds derived in validated methods, the global error estimates have no guarantees of correctness.

8   Conclusion and Future Work

We have presented a framework for developing *end-to-end* proofs verifying the accuracy and correctness of imperative implementations of ODE solvers for IVPs, and have demonstrated the utility of this framework on leapfrog integration of the simple harmonic oscillator. Our framework leverages several libraries and tools embedded in the Coq proof assistant to modularize the verification process. The end-to-end result is a proof that the floating-point solution returned by a C implementation of leapfrog integration of the harmonic oscillator is an accurate solution to the IVP. This proof is composed of two main theorems that clearly disentagle program correctness from numerical accuracy.

Our main theorem regarding the numerical accuracy of the program treats round-off error, discretization error, and global error propagation distinctly, and makes clear how discretization error can be used to derive tight bounds on round-off error. By treating each source of error in this modular way, our framework could be extended to include additional sources of error of concern in the solution to IVPs for ODEs, such as error in the data and uncertainty in the model; we leave this extension to future work.

In its current state, our framework would require substantial user effort in order to be re-used on a different IVP or ODE solver. This obstacle could be overcome by developing proof automation for each component of the error analysis presented in Sect. 3. In particular, the derivation of local discretization error presented in Sect. 3.1 is standard: given user-supplied input for the order at which to truncate the Taylor series expansion, the specification for the continuous system could be used to derive a maximum local error bound supposing that the autonomous IVP is Lipschitz continuous in x as discussed in Sect. 3. This assumption on the IVP is enough to guarantee existence and uniqueness of a solution [13,14]; we leave the Coq formalization of existence and uniqueness theorems (using Coquelicot) to future work. One could mostly automate the error propagation analysis in Sect. 3.2: if the user supplied (or if an unverified tool calculated) a transition matrix $M(h)$ for the ODE solver and a guess C for an upper bound on a suitable norm of $M(h)$, one would only need to discharge a proof that $\|M(h)\| \leq C$. Finally, while the proof of a local round-off error bound is already mostly automatic using VCFloat, employing the global discretization error bound in the proof of local round-off error is currently done by the user; this process could be completed in an automatic way.

Acknowledgments. This work benefited substantially from discussions with David Bindel. We thank Michael Soegtrop for his close reading and helpful feedback. Ariel Kellison is supported by the U.S. Department of Energy, Office of Science, Office of Advanced Scientific Computing Research, Department of Energy Computational Science Graduate Fellowship under Award Number DE-SC0021110.

References

1. Hairer, E., Lubich, C., Wanner, G.: Geometric numerical integration illustrated by the Störmer-Verlet method. Acta Numerica **12**, 399–450 (2003)
2. Nedialkov, N.S., Jackson, K.R., Corliss, G.F.: Validated solutions of initial value problems for ordinary differential equations. Appl. Math. Comput. **105**(1), 21–68 (1999)
3. Lin, Y., Stadtherr, M.A.: Validated solutions of initial value problems for parametric ODEs. Appl. Numer. Math. **57**(10), 1145–1162 (2007)
4. dit Sandretto, J.A., Chapoutot, A.: Validated explicit and implicit Runge-Kutta methods. Reliable Computing Electronic Edition, 22 July 2016
5. Rauh, A., Auer, E.: Verified simulation of ODEs and their solution. Reliab. Comput. **15**(4), 370–381 (2011)
6. Nedialkov, N.S., Jackson, K.R.: ODE software that computes guaranteed bounds on the solution. In: Langtangen, H.P., Bruaset, A.M., Quak, E. (eds.) Advances in Software Tools for Scientific Computing, pp. 197–224. Springer, Heidelberg (2000). https://doi.org/10.1007/978-3-642-57172-5_6
7. Nedialkov, N.S.: Interval tools for ODEs and DAEs. In: 12th GAMM - IMACS International Symposium on Scientific Computing, Computer Arithmetic and Validated Numerics (SCAN 2006), p. 4 (2006)
8. Appel, A.W.: Verified software toolchain. In: Barthe, G. (ed.) ESOP 2011. LNCS, vol. 6602, pp. 1–17. Springer, Heidelberg (2011). https://doi.org/10.1007/978-3-642-19718-5_1
9. Boldo, S., Lelay, C., Melquiond, G.: Coquelicot: a user-friendly library of real analysis for Coq. Math. Comput. Sci. **9**(1), 41–62 (2015)
10. Boldo, S., Melquiond, G.: Computer Arithmetic and Formal Proofs: Verifying Floating-point Algorithms with the Coq System. Elsevier, Amsterdam (2017)
11. Appel, A.W., Kellison, A.E.: VCFloat2: floating-point error analysis in Coq. Draft (2022)
12. Ramananandro, T., Mountcastle, P., Meister, B., Lethin, R.: A unified Coq framework for verifying C programs with floating-point computations. In: Proceedings of the 5th ACM SIGPLAN Conference on Certified Programs and Proofs, CPP 2016, pp. 15–26. Association for Computing Machinery, New York (2016)
13. Hairer, E., Norsett, S.P., Wanner, G.: Solving Ordinary Differential Equations I. Nonstiff Problems, 2nd rev. edition. Springer, Heidelberg (1993). https://doi.org/10.1007/978-3-540-78862-1. Corr. 3rd printing edition, 1993
14. LeVeque, R.J.: Finite Difference Methods for Ordinary and Partial Differential Equations. Society for Industrial and Applied Mathematics, Philadelphia (2007)
15. Hairer, E., Lubich, C., Wanner, G.: Geometric Numerical Integration. Structure-Preserving Algorithms for Ordinary Differential Equations. Springer Series in Computational Mathematics, vol. 31, 2nd edn. Springer, Heidelberg (2006). https://doi.org/10.1007/3-540-30666-8
16. Bou-Rabee, N., Sanz-Serna, J.M.: Geometric integrators and the Hamiltonian Monte Carlo method. Acta Numerica **27**, 113–206 (2018)
17. Blanes, S., Casas, F., Sanz-Serna, J.M.: Numerical integrators for the hybrid Monte Carlo method. SIAM J. Sci. Comput. **36**(4), A1556–A1580 (2014)
18. Skeel, R.D.: Integration schemes for molecular dynamics and related applications. In: Ainsworth, M., Levesley, J., Marletta, M. (eds.) The Graduate Student's Guide to Numerical Analysis '98. Springer, Heidelberg (1999). https://doi.org/10.1007/978-3-662-03972-4_4

19. Appel, A.W., et al.: Program Logics for Certified Compilers. Cambridge University Press, Cambridge (2014)

20. Beringer, L., Appel, A.W.: Abstraction and subsumption in modular verification of C programs. Formal Methods Syst. Des. **58**, 322–345 (2021). https://doi.org/10.1007/s10703-020-00353-1

21. Boldo, S., Clément, F., Filliâtre, J.-C., Mayero, M., Melquiond, G., Weis, P.: Trusting computations: a mechanized proof from partial differential equations to actual program. Comput. Math. Appl. **68**(3), 325–352 (2014)

22. Boldo, S., Faissole, F., Chapoutot, A.: Round-off error analysis of explicit one-step numerical integration methods. In: 24th IEEE Symposium on Computer Arithmetic, London, UK, July 2017

23. Daumas, M., Melquiond, G.: Certification of bounds on expressions involving rounded operators. ACM Trans. Math. Softw. **37**(1), 1–20 (2010)

24. de Dinechin, F., Lauter, C., Melquiond, G.: Certifying the floating-point implementation of an elementary function using Gappa. IEEE Trans. Comput. **60**(2), 242–253 (2011)

25. Immler, F., Hölzl, J.: Numerical analysis of ordinary differential equations in Isabelle/HOL. In: Beringer, L., Felty, A. (eds.) ITP 2012. LNCS, vol. 7406, pp. 377–392. Springer, Heidelberg (2012). https://doi.org/10.1007/978-3-642-32347-8_26

26. Corliss, G.F.: Guaranteed Error Bounds for Ordinary Differential Equations. Oxford University Press, Oxford (1994)

27. Nedialkov, N.S., Jackson, K.R., Pryce, J.D.: An effective high-order interval method for validating existence and uniqueness of the solution of an IVP for an ODE. Reliab. Comput. **7**(6), 449–465 (2001)

28. Jackson, K.R., Nedialkov, N.S.: Some recent advances in validated methods for IVPs for ODEs. Appl. Numer. Math. **42**(1), 269–284 (2002)

29. Rihm, R.: Interval methods for initial value problems in ODEs. In: Topics in Validated Computations: Proceedings of IMACS-GAMM International Workshop on Validated Computation, September 1993

30. Shampine, L.F.: Error estimation and control for ODEs. J. Sci. Comput. **25**(1), 3–16 (2005)

31. Cao, Y., Petzold, L.: A posteriori error estimation and global error control for ordinary differential equations by the adjoint method. SIAM J. Sci. Comput. **26**(2), 359–374 (2004)

32. Kehlet, B., Logg, A.: A posteriori error analysis of round-off errors in the numerical solution of ordinary differential equations. Numer. Algorithms **76**(1), 191–210 (2017)

Neural Network Precision Tuning Using Stochastic Arithmetic

Quentin Ferro[1]($\boxtimes$), Stef Graillat[1], Thibault Hilaire[1], Fabienne Jézéquel[1,2], and Basile Lewandowski[1]

[1] Sorbonne Université, CNRS, LIP6, 75005 Paris, France
{quentin.ferro,stef.graillat,thibault.hilaire,fabienne.jezequel}@lip6.fr
[2] Université Paris-Panthéon-Assas, 75005 Paris, France

Abstract. Neural networks can be costly in terms of memory and execution time. Reducing their cost has become an objective, especially when integrated in an embedded system with limited resources. A possible solution consists in reducing the precision of their neurons parameters. In this article, we present how to use auto-tuning on neural networks to lower their precision while keeping an accurate output. To do so, we use a floating-point auto-tuning tool on different kinds of neural networks. We show that, to some extent, we can lower the precision of several neural network parameters without compromising the accuracy requirement.

Keywords: Precision · Neural networks · Auto-tuning · Floating-point · Stochastic arithmetic

1 Introduction

Neural networks are nowadays massively used and are becoming larger and larger. They often need a lot of resources, which can be a problem, especially when used in a critical embedded system with limited computing power and memory. Therefore it can be very beneficial to optimise the numerical formats used in a neural network. This article describes how to perform precision auto-tuning of neural networks. From a neural network application and an accuracy requirement on its results, it is shown how to obtain a mixed precision version using the PROMISE tool [15]. A particularity of PROMISE is the fact that it uses stochastic arithmetic [38] to control rounding errors in the programs it provides.

Minimizing the format of variables in a numerical simulation can offer advantages in terms of execution time, volume of data exchanged and energy consumption. During the past years several algorithms and tools have been proposed for precision auto-tuning. On the one hand, tools such as FPTuner [8], Salsa [10], Rosa/Daisy [11,12], TAFFO [5], POP [2] rely on a static approach and are not intended to be used on very large code. On the other hand, dynamic tools such as CRAFT HPC [24], Precimonious [33], HiFPTuner [16], ADAPT [30], FloatSmith [25], PROMISE [15], have been proposed for precision auto-tuning

O. Isac et al. (Eds.): NSV 2022/FoMLAS 2022, LNCS 13466, pp. 164–186, 2022.
https://doi.org/10.1007/978-3-031-21222-2_10

in large HPC code. Moreover, tools have been recently developed for precision auto-tuning on GPUs: AMPT-GA [22], GPUMixer [23], GRAM [18]. A specificity of PROMISE lies in the fact that it provides mixed precision programs validated owing to stochastic arithmetic [38], whereas other dynamic tools rely on a reference result possibly affected by rounding errors. PROMISE has been used in various applications based on linear algebra kernels, but not yet for precision auto-tuning in deep neural networks.

While much effort has been devoted to the safety and robustness of deep learning code (see for instance [13,27,28,32,34]) a few studies have been carried out on the effects of rounding error propagation on neural networks. Verifiers such as MIPVerify [36] are designed to check properties of neural networks and measure their robustness. However, the impact of floating-point arithmetic both on neural networks and on verifiers is pointed out in [42]. Because of rounding errors, the actual robustness and the robustness provided by a verifier may radically differ. In [9,41] it is shown how to control the robustness of different neural networks with greater efficiency using interval arithmetic based algorithms. In [26] a software framework is presented for semi-automatic floating-point error analysis in the inference phase of deep neural networks. This analysis provides absolute and relative error bounds owing to interval and affine arithmetics.

Precision tuning of neural networks using fixed-point arithmetic has been studied in [3]. Owing to the solution of a system of linear contraints, the fixed-point precision of each neuron is determined, taking into account a certain error threshold.

In this article, we consider floating-point precision tuning, which is also studied in [20]. Focusing on interpolator networks, i.e. networks computing mathematical functions, the authors propose an algorithm that takes into account a given tolerance δ on the relative error between the assumed correctly computed function and the function computed by the network. The main difference with the present article is the auto-tuning algorithm: in [20] the precision is optimized by solving a linear programming problem, while PROMISE uses a hypothesis-trial-result approach through the Delta-Debug algorithm [40]. Furthermore, the algorithm in [20] relies on a reference result that may be altered by rounding errors, while PROMISE uses stochastic arithmetic for the numerical validation of its results.

Stochastic arithmetic uses for rounding error estimation a random rounding mode: the result of each arithmetic operation is rounded up or down with the same probability. As a remark, another stochastic rounding often used in neural network training and inference uses a probabiliy that depends on the position of the exact result with respect to the rounded ones (see for instance [14,17,29, 31,35,39]). This stochastic rounding does not aim at estimating rounding errors, it enables the update of small parameters and avoids stagnations that may be observed with round to nearest.

In this work, we consider tuning the precision of an already trained neural network. One of our contributions is a methodology for tuning the precision of a neural network using PROMISE in order to obtain the lowest precision for each of its parameters, while keeping a certain accuracy on its results. We present and compare the results obtained for different neural networks: an approximation of the sine function, an image classifier processing the MNIST dataset (2D pictures of handwritten digits), another image classifier using this time convolutional layers and processing the CIFAR10 dataset (3D images of different classes), and the last one introduced in [4] and used in [26] that aims at approximating a Lyapunov function of a nonlinear controller for an inverted pendulum.

After a preliminary reminder on deep neural networks and stochastic arithmetic in Sect. 2, Sect. 3 describes our methodology and Sect. 4 presents our results considering the different neural networks previously mentioned.

2   Preliminary

2.1   Neural Networks

An artificial neural network is a computing system defined by several neurons distributed on different layers. Generally, we consider dense layers that take as an input a vector and in which the main computation is a matrix-vector product. In this case, from one layer to another, a vector of neurons $x^{(k)} \in \mathbb{R}^{n_k}$ with $k \in \mathbb{N}$ is transformed into a vector $x^{(k+1)} \in \mathbb{R}^{n_{k+1}}$ by the following equation

$$x^{(k+1)} = g^{(k)}(W^{(k)}x^{(k)} + b^{(k)}) \tag{1}$$

where $W^{(k)} \in \mathbb{R}^{n_{k+1} \times n_k}$ is a weight matrix, $b^{(k)} \in \mathbb{R}^{n_{k+1}}$ a bias vector and $g^{(k)}$ an activation function. The activation function is a non-linear and often monotonous function. Most common activation functions are described below.

- Sigmoid: computes $\sigma(x) = 1/(1 + e^{-x}) \ \forall x \in \mathbb{R}$
- Hyperbolic Tangent: applies $\tanh(x) \ \forall x \in \mathbb{R}$
- Rectified Linear Unit: $\text{ReLU}(x) = \max(x, 0) \ \forall x \in \mathbb{R}$
- Softmax: normalizes an input vector x into a probability distribution over the output classes. For each element x_i in x, $\text{softmax}(x_i) = e^{x_i} / \sum e^{x_j}$

Figure 1 shows a graphical representation of a neural network with two layers.

Dense layers are sometimes generalized to multidimensional arrays and involve tensor products. Other layers exist such as convolution layers often used on multi-dimensional arrays to extract features out of the data. To do so, a convolution kernel is applied to the input data to produce the output. Different kinds of layers, for instance pooling layers and flatten layers, do not require weight nor bias. Pooling layers reduce the size of the input data by summarizing it by zones given a function such as the maximum or the average. Flatten layers are intended to change the shape of data, from a 3D tensor to a vector for example and can be used to pass from a convolutional layer to a dense layer.

Input   Layer 1   Layer 2

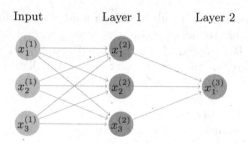

Fig. 1. Neural network with two layers

2.2 Floating-Point Arithmetic

A floating-point number x in base β is defined by:

$$x = (-1)^s \times m \times \beta^e \tag{2}$$

with s its sign being either 1 or 0, m its significand being an integer and e its exponent being also an integer. In this paper, we consider binary floating-point numbers, i.e. numbers in base $\beta = 2$ that adhere to the IEEE 754 Standard [1]. The IEEE 754 Standard defines different formats with a fixed number of bits for the significand and the exponent. The number of bits for the significand is the precision p, hence the significand can take values ranging from 0 to $\beta^p - 1$. The exponent e ranges from e_{min} to e_{max} with $e_{min} = 1 - e_{max}$ and $e_{max} = 2^{len(e)-1} - 1$, with $len(e)$ the exponent length in bits. The sizes of the three different formats used in this paper, commonly named half, single (or float), and double, are sum up in Table 1. As a remark, another 16-bit format called bfloat16 exists, for example on ARM NEON CPUs. Owing to their 8-bit-large exponent, bfloat16 numbers benefit from a wider range, but have a lower significand precision (8 bits instead of 11).

Table 1. Basic binary IEEE 754 formats

Name	Format	Length	Sign	Significand length[a]	Exponent length
Half	binary16	16 bits	1 bit	11 bits	5 bits
Single	binary32	32 bits	1 bit	24 bits	8 bits
Double	binary64	64 bits	1 bit	53 bits	11 bits

[a] Including the implicit bit (which always equals 1 for normal numbers, and 0 for subnormal numbers). The implicit bit is not stored in memory.

2.3 Discrete Stochastic Arithmetic (DSA)

Discrete Stochastic Arithmetic (DSA) is a method for rounding error analysis based on the CESTAC method [7,37]. The CESTAC method allows the estimation of round-off error propagation that occurs when computing with floating-point numbers. Based on a probabilistic approach, it uses a random rounding mode: at each operation, the result is either rounded up or down with the same probability. Using this rounding mode, the same program is run N times giving us N samples $R_1, \ldots, R_N$ of the computed result R. The accuracy of the computed result (i.e. its number of exact significant digits) can be estimated using Student's law with the confidence level 95%. In practice the sample size is $N = 3$. Indeed, it has been shown that $N = 3$ is in some reasonable sense the optimal value. The estimation with $N = 3$ is more reliable than with $N = 2$ and increasing the sample size does not improve the quality of the estimation. Theoretical elements can be found in [6, 37].

The CADNA [6, 21, 37] (Control of Accuracy and Debugging for Numerical Applications) software[1] implements DSA in code written in C, C++ or Fortran. It introduces new variable types, the stochastic types. Each stochastic variable contains three floating-point values and one integer being the exact number of correct digits. CADNA can print each computed value with only its exact significant digits. In practice, owing to operator overloading, the use of CADNA only requires to change declaration of variables and input/output statements.

2.4   The PROMISE Software

The PROMISE software[2] aims at reducing the precision of the variables in a given program. From an initial code and a required accuracy, it returns a mixed precision code, lowering the precision of the different variables while keeping a result that satisfies the accuracy constraint. To do so, some variables are declared as custom typed variables that PROMISE recognizes. PROMISE will consider tweaking their precisions. Different variables can be forced to have the same precision by giving them the same custom type. It may be useful to avoid compilation errors or casts of variables.

PROMISE computes a reference result using CADNA and relies on the Delta-Debug algorithm [40] to test different type configurations, until a suitable one lowering the precision while satisfying the accuracy requirement is found. PROMISE provides a transformed program that can mix half, single and double precision. Half precision can be either native on CPUs that support it or emulated using a library developed by C. Rau[3]. PROMISE dataflow is presented in Fig. 2. After computing the reference result, PROMISE tries to lower the precision of the variables from double to single, then from single to half, using twice the Delta-Debug algorithm. The accuracy requirement may concern one or several variables (e.g. in an array). PROMISE checks that the number of common digits between the computed result(s) and the reference result(s) is at least the required accuracy. In the case of several variables, the requirement has to be fulfilled by all of them.

---

[1] http://cadna.lip6.fr.

[2] http://promise.lip6.fr.

[3] http://half.sourceforge.net.

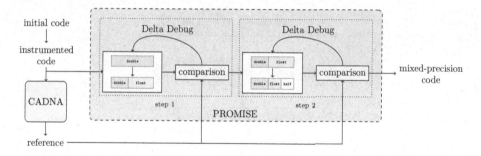

Fig. 2. PROMISE dataflow

3 Methodology

For neural network management, we use Python language with either Keras[4] or PyTorch[5]. Keras and Pytorch are two open-source Python libraries that implement structures and functions to create and train neural network models. Both of them also allow us to save our model in HDF5 (Hierarchical Data Format)[6], a file format designed to store and organize large data. HDF5 uses only two types of objects: datasets that are multidimensional arrays of homogeneous type, and groups, which contain datasets or other groups. HDF5 files can be read by Python programs using the h5py package. The associated data can be manipulated with Pandas[7], a Python library that proposes data structures and operations to manage large amount of data.

Keras is used to develop, train and save our neural network models, except in the case of the inverted pendulum which uses PyTorch. As already mentioned in the introduction, precision tuning is performed on trained models, hence in the inference stage. The process path is summarized in Fig. 3. For each neural network, we first convert the HDF file to CSV files using a Python script. The script loads the HDF file, stores the parameters in Pandas DataFrames, and then saves the parameters in CSV files using the Pandas DataFrame *to_csv* function. For each layer that needs it, we create a CSV file with the weights of the layer and a CSV file with the bias of the layer. Indeed, some layers do not need weights nor bias, for example flattening layers that only change the data shape (from 2 dimensions to 1 dimension for example). Secondly, we use the data in the CSV files to create a C++ program, once again using a Python script that reads the CSV files and creates the necessary variables and computation. The translation scripts are based on the work done in the keras2c[8] library. Once the C++ version created, we apply PROMISE on it.

---

[4] https://keras.io.
[5] https://pytorch.org.
[6] https://www.hdfgroup.org.
[7] https://pandas.pydata.org.
[8] https://f0uriest.github.io/keras2c/.

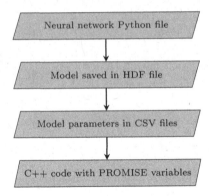

Fig. 3. Flowchart of the translation from a Python neural network to a C++ program with PROMISE variables

4   Experimental Results

Results obtained for four different neural networks are presented in this section. For the neural networks using a database, test_data[i] refers to $(i+1)^{th}$ test input provided by the database. PROMISE is applied to each neural network considering one type by neuron (half, single or double), then one type per layer, i.e. all the parameters of a layer have the same precision. In our analysis, the difference between the two approaches lies in the number of different type declarations in the code. However, it must be pointed out that, in dense layers, having one type per neuron implies independent dot products, whereas having one type per layer would enable one to compute matrix-vector products that could perform better.

In our experiments, the input is in double precision. In accordance with Fig. 2, for any neural network, the reference value is the value computed at the very first step of PROMISE. All the results presented in this section have been obtained on a 2.80 GHz Intel Core i5-8400 CPU having 6 cores with 16 GB RAM except indicated otherwise.

4.1   Sine Neural Network

To approximate the sine function, we use a classical densely-connected neural network with 3 layers. It is a toy problem, since using a neural network to compute sine is not necessary. However, this simple example validates our approach. The tanh activation function is used in the 3 different dense layers. The layers have respectively 20, 6 and 1 neuron(s) and the input is a scalar value x. Figure 4 presents the computation carried out by the neural network, considering one type per neuron. Colored variables are PROMISE variables, the precision of which can be tweaked. Variables with the same color have the same precision. The parameters of a neuron (weight(s) and bias) have the same color, hence the same type. The output type of each layer is also tuned. In this example, we assign types to 27 neurons and 3 outputs.

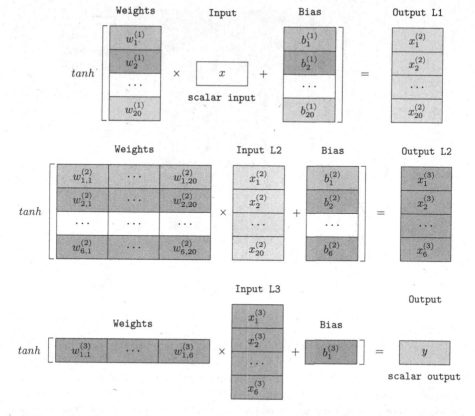

Fig. 4. Computation carried out in the sine neural network

Figure 5 displays the distribution of the different types with input value 0.5 considering one type per neuron. The x-axis presents the required accuracy on the results, i.e. the number of significant digts in common with the reference result computed using CADNA. We can notice the evolution of the distribution depending on the number of exact significant digits required on the result. As expected, first we only have half precision variables, then some of them start to be in single, then in double precision, until eventually all of them are in double precision. Therefore, requiring the highest accuracy is not compatible with lowering the precision in this neural network. But still, a good compromise can be found, since we only have single and half precision variables for a required accuracy up to 7 digits, and still have 1/3 of single precision variables for a required accuracy up to 9 digits.

Figure 5 also presents the computation time of PROMISE in seconds for each required accuracy. It consists of the time to compute the reference result, and the time to apply the Delta-Debug algorithm twice (from double to single precision then from single to half precision), compiling and executing the tested distribution each time. It can be noticed that the computation time (less than 2 min) remains reasonable given the 3^{30} possibilities.

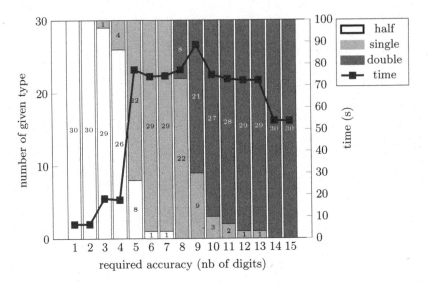

Fig. 5. Number of variables of each type and computation time for the sine neural network with input value 0.5

Figure 6 shows the type distribution considering one precision per layer. This approach per layer enables one to decrease the execution time of PROMISE, but it does not really help lowering the precision of the network parameters. Indeed, each time a parameter in a layer requires a higher precision, all the parameters of the same layer pass in higher precision. But still, it can be noticed that the first layer (that represents 2/3 of the neurons) stays in half precision for a required accuracy up to 4 digits.

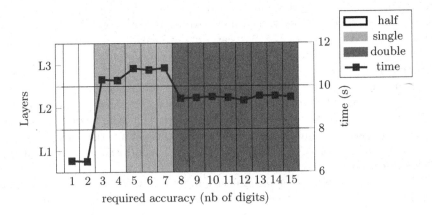

Fig. 6. Precision of each layer for the sine neural network with input value 0.5

Figures in Appendix A.1 and Appendix A.2 show that the input value can have a slight impact on the type distribution. We compare the results for two input values randomly chosen: 0.5 and 2.37. From Fig. 5, with input value 0.5, PROMISE provides a type distribution with 8 half precision neurons and 22 single precision neurons for a required accuracy of 5 digits. From Appendix A.1, with input value 2.37, only 4 half precision neurons are obtained for a required accuracy of 5 digits. Actually, if 3, 4, or 5 digits are required, less neurons are in half precision than with input 0.5. But if 8 digits are required, one neuron remains in half precision with input 2.37, while no neuron is in half precision with input 0.5. Nevertheless, the type distribution with respect to the required accuracy remains globally the same, parameters all starting as half precision variables, then passing to higher precision.

4.2 MNIST Neural Network

Experiments have been carried out with an image classification neural network processing MNIST data of handwritten digits[9]. This neural network also uses classical dense layers. The main difference in this case is that the entry is a vector of size 784 (flatten image) and the output is a vector of size 10. This neural network consists of two layers: the first one with 64 neurons and the activation function ReLU, and the second one with 10 neurons and the activation function softmax which provides the probability distribution for the 10 different classes. Considering as previously one type per neuron, plus one type for the output of each layer, 76 different types have to be set either to half, single, or double precision.

Figure 7 shows the type distribution considering one type per neuron with the input image test_data[61]: the 62nd test data out of the 10,000 provided by MNIST. The x-axis represents the required accuracy on the output consisting in a vector of size 10. The maximum accuracy on the output is 13 digits, higher expectation could not be matched. Such high accuracy is nonetheless not realistic because not necessary for a classifier. However, exhaustive tests have been performed: all possible accuracies in double precision have been successively required.

The main difference with the sine neural network lies in the fact that a significant number of neurons stay in half precision no matters the required accuracy. Depending on the input, around 50% of neurons can stay in half precision and sometimes nearly 60% as shown in Appendix B.2. Thus, applying PROMISE to this neural network, even when requiring the highest accuracy, can help lowering the precision of its parameters. Neurons that keep the lowest precision are not the same depending on the input, but they always belong to the first layer. Hence, the first layer seems to have less impact on the output accuracy than the second one.

---

[9] http://yann.lecun.com/exdb/mnist.

The computation times also reported in Fig. 7 are much higher than for the sine approximation network. For the majority of the accuracy requirements, more than 15 min are necessary to obtain a mixed precision version of the neural network. The MNIST neural network has one layer less than the sine approximation network, but more neurons. Since we consider one precision per neuron, the number of possible type configurations (3^{76}) is much higher, hence the computation time difference. However, the execution time of PROMISE remains reasonable and performing such a tuning by hand would have been much more time consuming.

With both the sine neural network and MNIST neural network, PROMISE execution time tends to increase with the accuracy requirement. This can be explained by the Delta-Debug algorithm in PROMISE. As previously described, PROMISE firstly checks whether the accuracy requirement can be satisfied with double precision. Then, owing to the Delta-Debug algorithm, PROMISE tries to lower the precision of most variables from double to single, and this can be very fast if single precision is enough to match the required accuracy. Finally, PROMISE tries to transform the single precision declarations into half precision ones, and this can be fast if half precision is suitable for all these declarations. The number of programs compiled and executed by PROMISE tends to increase with the required accuracy on the results. For instance, in the case of MNIST neural network, if 1 or 2 digits are required, 18 type configurations are tested by PROMISE, whereas if 7 digits are required, 260 configurations are tested.

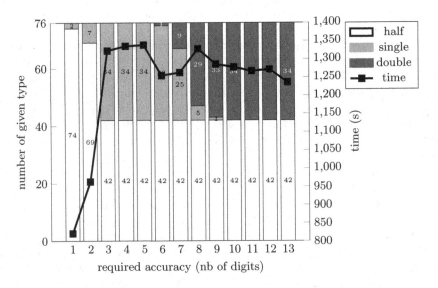

Fig. 7. Number of variables of each type and computation time for MNIST neural network with test_data[61] input

Results obtained considering one precision per layer are presented in Appendix B.1. The analysis is similar to the one previously given for the sine neural network. With the approach per layer, the execution time of PROMISE is lower than with the approach per neuron. But this approach forces some variables to be declared in higher precision. It can be noticed that with the approach per layer, both layers have the same precision. All the network parameters share the same type. Appendix B.2 and Appendix B.3 present the results obtained with another input. Like with the sine neural network, changing the input induces slight changes in the type configurations provided by PROMISE. However, the same trend can be observed.

4.3 CIFAR Neural Network

The neural network considered here is also an image classifier, but this time processing the CIFAR10 dataset. CIFAR[10] is a dataset having 100 classes of colored images and the CIFAR10 dataset is reduced to 10 classes. Because images are of size $32 \times 32 \times 3$ (32 pixels width, 32 pixels height, 3 color channels), the network input is a 3D tensor of shape $(32, 32, 3)$. The neural network consists of 5 layers: a convolutional layer having 32 neurons with activation function ReLU, followed by a max pooling of size 2×2, a convolutional layer having 64 neurons with activation function ReLU, a flatten layer, and finally a dense layer of 10 neurons with activation function softmax. Taking into account one type per neuron and one type for each layer output, 111 types can be set.

Results presented here have been obtained on a 2.80 GHz Intel Core i9-10900 CPU having 20 cores and 64GB RAM. The maximum possible accuracy on the results is 13 digits. Although such a high accuracy is not necessary in a classifier network, exhaustive tests have been performed, like for the MNIST neural network. Results reported here refer to two input images out of the 10,000 provided by CIFAR10. Figure 8 and Appendix C.2 present the type configurations given by PROMISE with respectively test_data[386] and test_data[731], considering one type per neuron. Appendix C.1 and Appendix C.3 show the results obtained with the same input images considering one type per layer.

PROMISE computation time tends to increase with the required accuracy. As already mentioned in Sect. 4.2, this can be explained by the Delta Debug algorithm. As previously observed, considering one type per layer results in lower PROMISE computation times, but often in uniform precision programs. Again, the input image slightly impacts the type configurations provided by PROMISE and the same trend can be observed.

---

[10] https://www.cs.toronto.edu/~kriz/cifar.html.

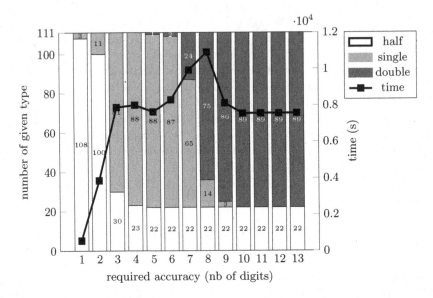

Fig. 8. Number of variables of each type and computation time for CIFAR neural network with test_data[386] input

Experiments have been carried out with neural networks also processing CIFAR10, but with more layers (up to 8 layers). Again, PROMISE could provide suitable type configurations taking into account accuracy requirements. However, PROMISE execution (that includes the compilation and execution of various programs) makes exhaustive tests more difficult with such neural networks. Possible PROMISE improvements described in Sect. 5 would enable precision tuning in larger neural networks that are themselves time consuming.

4.4   Inverted Pendulum

We present here results obtained with a neural network introduced in [4] in the context of reinforcement learning for autonomous control. In [4] methods are proposed for certified approximation of neural controllers and this neural network, related to an inverted pendulum, is used in a program that provides an approximation of a Lyapunov function. This neural network consists of 2 dense layers and uses the tanh activation function. The input is a state vector $x \in \mathbb{R}^2$ and the output, a scalar value in $\mathbb{R}$ is an approximated value of the Lyapunov function for the state vector input. The first layer has 6 neurons and the second layer only one. Given the proximity between the two neural network models, results are expected to be close to the ones obtained for the sine approximation.

Figure 9 presents both the distribution of the different precisions and the execution time of PROMISE with respect to the accuracy requirement for input $(0.5, 0.5)$. We consider here one type per neuron. As expected, the trend observed for the type configurations is the same as with the sine approximation. As the required accuracy increases, the precision of the network parameters also increases. If one digit is required, all the parameters can be declared in half precision, and if at least 11 digits are required all the parameters must be in double precision. The computation time remains reasonable whatever the required accuracy. As previously observed, the computation time tends to increase with the required accuracy because of the number of type configurations tested by PROMISE.

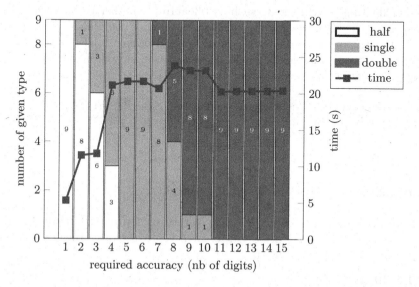

Fig. 9. Number of variables of each type and computation time for the pendulum neural network with input $(0.5, 0.5)$

Appendix D.1 presents the results obtained considering one type per layer. Again, with this approach, the execution time of PROMISE is lower, but most configurations are actually in uniform precision. Results in Appendix D.2 and Appendix D.3 refer to an input consisting of two negative values $(-3, -6)$. As previously observed, changing the input induces no significant difference.

5   Conclusion and Perspectives

We have shown with different kinds of neural networks having different types of layers how to lower the precision of their parameters while still satisfying a desired accuracy. Considering one type per neuron, mixed precision programs can be provided by PROMISE. Considering one type per layer enables one to reduce PROMISE execution time, however this approach often leads to uniform precision programs. It has been observed that with both approaches input values have actually a low impact on the type configurations obtained.

We plan to analyse the execution time of the mixed precision programs obtained with PROMISE on a processor with native half precision. Other perspectives consist in improving PROMISE. Another accuracy test more adapted to image classification networks could be proposed. We could improve the Delta-Debug algorithm used in PROMISE. Optimizations of the Delta-Debug algorithm are described in [19], including the parallelization potential. We could also consider the parallelization of PROMISE itself, i.e. applying PROMISE to different parts of a code in parallel. The extension of PROMISE and CADNA to other floating-point formats such as bfloat16 is another perspective. Taking benefit from the GPU version of CADNA, PROMISE could also be extended to GPUs. Floating-point auto-tuning in arbitrary precision is also a possible perspective that would enable the automatic generation of programs with a suitable type configuration for architectures such as FPGAs.

Acknowledgements. This work was supported by the InterFLOP (ANR-20-CE46-0009) project of the French National Agency for Research (ANR).

Appendices

Appendix A   Sine Neural Network

Appendix A.1   Type Distribution for Sine Approximation with Input Value 2.37 Considering One Type per Neuron

See Fig. 10.

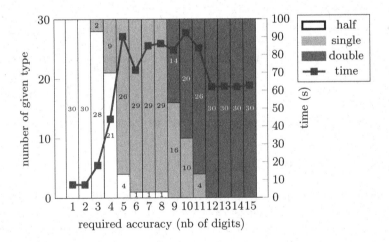

Fig. 10. Number of variables of each type and computation time for the sine neural network with input value 2.37

Appendix A.2 Type Distribution for Sine Approximation with Input Value 2.37 Considering One Type per Layer

See Fig. 11.

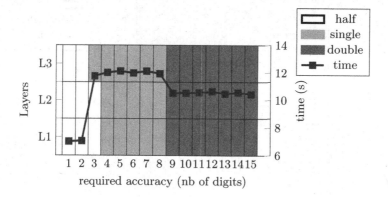

Fig. 11. Precision of each layer for the sine neural network with input value 2.37

Appendix B MNIST Neural Network

Appendix B.1 Type Distribution for MNIST with Test_data[61] Input Considering One Type per Layer

See Fig. 12.

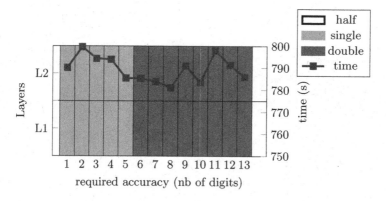

Fig. 12. Precision of each layer for MNIST neural network with test_data[61] input

Appendix B.2   Type Distribution for MNIST with Test_data[91] Input Considering one type per neuron

See Fig. 13.

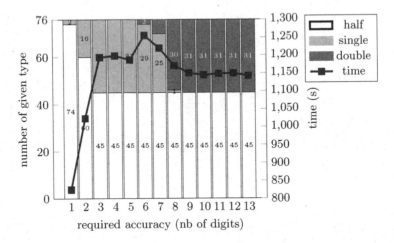

Fig. 13. Number of variables of each type and computation time for MNIST neural network with test_data[91] input

Appendix B.3   Type Distribution for MNIST with Test_data[91] Input Considering One Type per Layer

See Fig. 14.

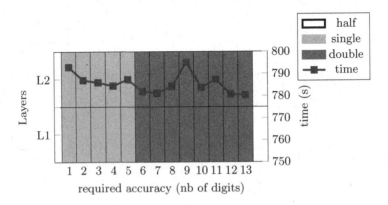

Fig. 14. Precision of each layer for MNIST neural network with test_data[91] input

Appendix C CIFAR Neural Network

Appendix C.1 Type Distribution for CIFAR with Test_data[386] Input Considering One Type per Layer

See Fig. 15.

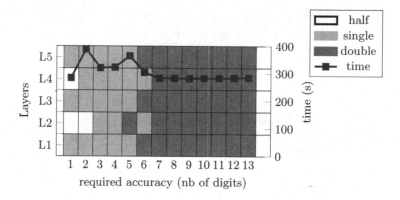

Fig. 15. Precision of each layer for CIFAR neural network with test_data[386] input

Appendix C.2 Type Distribution for CIFAR with Test_data[731] Input Considering One Type per Neuron

See Fig. 16.

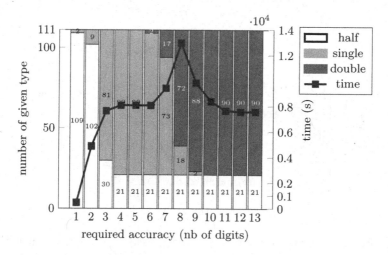

Fig. 16. Number of variables of each type and computation time for CIFAR neural network with test_data[731] input

Appendix C.3   Type Distribution for CIFAR with Test_data[731] Input Considering One Type per Layer

See Fig. 17.

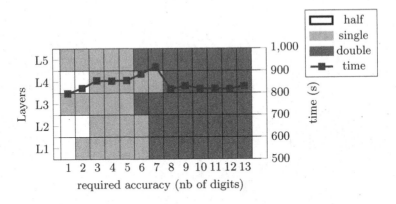

Fig. 17. Precision of each layer for CIFAR neural network with test_data[731] input

Appendix D   Inverted pendulum

Appendix D.1   Type Distribution for the Inverted Pendulum with Input (0.5,0.5) Considering One Type per Layer

See Fig. 18.

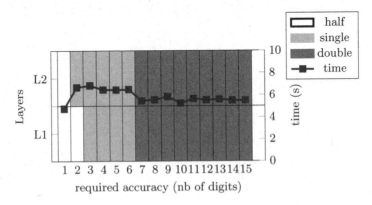

Fig. 18. Precision of each layer for the pendulum neural network with input (0.5, 0.5)

Appendix D.2 Type Distribution for the Inverted Pendulum with Input $(-3, -6)$ Considering One Type per Neuron

See Fig. 19.

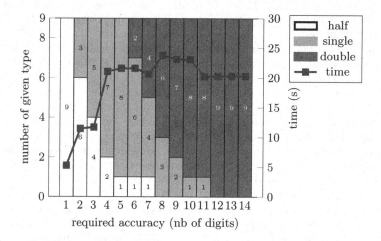

Fig. 19. Number of variables of each type and computation time for the pendulum neural network with input $(-3, -6)$

Appendix D.3 Type Distribution for the Inverted Pendulum with Input $(-3, -6)$ Considering One Type per Layer

See Fig. 20.

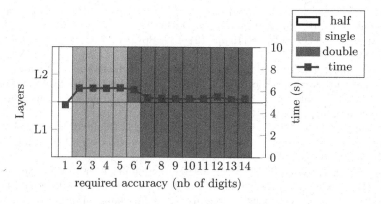

Fig. 20. Precision of each layer for the pendulum neural network with input $(-3, -6)$

References

1. IEEE Standard for Floating-Point Arithmetic, IEEE Std 754-2019 (Revision of IEEE 754-2008), pp. 1–84 (2019)
2. Adjé, A., Khalifa, D.B., Martel, M.: Fast and efficient bit-level precision tuning. arXiv:2103.05241 [cs] (2021)
3. Benmaghnia, H., Martel, M., Seladji, Y.: Fixed-point code synthesis for neural networks. Artificial Intelligence, Soft Computing and Applications, pp. 11–30 (2022). arXiv:2202.02095
4. Chang, Y.-C., Roohi, N., Gao, S.: Neural Lyapunov control. In: 33rd Conference on Neural Information Processing Systems (NeurIPS 2019) (2020). arXiv:2005.00611
5. Cherubin, S., Cattaneo, D., Chiari, M., Bello, A.D., Agosta, G.: TAFFO: tuning assistant for floating to fixed point optimization. IEEE Embed. Syst. Lett. **12**, 5–8 (2020)
6. Chesneaux, J.-M.: L'arithmétique stochastique et le logiciel CADNA, Habilitation à diriger des recherches. Université Pierre et Marie Curie, Paris, France (1995)
7. Chesneaux, J.-M., Graillat, S., Jezequel, F.: Numerical validation and assessment of numerical accuracy. Oxford e-Research Center (2009)
8. Chiang, W.-F., Baranowski, M., Briggs, I., Solovyev, A., Gopalakrishnan, G., Rakamarić, Z.: Rigorous floating-point mixed-precision tuning. In: Proceedings of the 44th ACM SIGPLAN Symposium on Principles of Programming Languages, POPL 2017, pp. 300–315. ACM, New York (2017)
9. Csendes, T.: Adversarial example free zones for specific inputs and neural networks. In: Proceedings of ICAI, pp. 76–84 (2020)
10. Damouche, N., Martel, M.: Mixed precision tuning with Salsa. In: Proceedings of the 8th International Joint Conference on Pervasive and Embedded Computing and Communication Systems, Porto, Portugal, pp. 47–56. SCITEPRESS - Science and Technology Publications (2018)
11. Darulova, E., Izycheva, A., Nasir, F., Ritter, F., Becker, H., Bastian, R.: Daisy - framework for analysis and optimization of numerical programs (tool paper). In: Beyer, D., Huisman, M. (eds.) TACAS 2018. LNCS, vol. 10805, pp. 270–287. Springer, Cham (2018). https://doi.org/10.1007/978-3-319-89960-2_15
12. Darulova, E., Kuncak, V.: Towards a compiler for reals. ACM Trans. Programm. Lang. Syst. (TOPLAS) **39**, 8:1–8:28 (2017)
13. Dutta, S., Jha, S., Sankaranarayanan, S., Tiwari, A.: Output range analysis for deep feedforward neural networks. In: Dutle, A., Muñoz, C., Narkawicz, A. (eds.) NFM 2018. LNCS, vol. 10811, pp. 121–138. Springer, Cham (2018). https://doi.org/10.1007/978-3-319-77935-5_9
14. Essam, M., Tang, T.B., Ho, E.T.W., Chen, H.: Dynamic point stochastic rounding algorithm for limited precision arithmetic in deep belief network training. In: 2017 8th International IEEE/EMBS Conference on Neural Engineering (NER), Shanghai, China, pp. 629–632. IEEE, May 2017
15. Graillat, S., Jézéquel, F., Picot, R., Févotte, F., Lathuilière, B.: Auto-tuning for floating-point precision with discrete stochastic arithmetic. J. Comput. Sci. **36**, 101017 (2019)
16. Guo, H., Rubio-González, C.: Exploiting community structure for floating-point precision tuning. In: Proceedings of the 27th ACM SIGSOFT International Symposium on Software Testing and Analysis, Amsterdam Netherlands, pp. 333–343. ACM, July 2018

17. Gupta, S., Agrawal, A., Gopalakrishnan, K., Narayanan, P.: Deep learning with limited numerical precision. In: Proceedings of the 32nd International Conference on International Conference on Machine Learning - Volume 37, ICML 2015 (2015). arXiv:1502.02551

18. Ho, N.-M., Silva, H.D., Wong, W.-F.: GRAM: a framework for dynamically mixing precisions in GPU applications. ACM Trans. Archit. Code Optim. **18**, 1–24 (2021)

19. Hodován, R., Kiss, Á.: Practical improvements to the minimizing delta debugging algorithm. In: Proceedings of the 11th International Joint Conference on Software Technologies, Lisbon, Portugal, pp. 241–248. SCITEPRESS - Science and Technology Publications (2016)

20. Ioualalen, A., Martel, M.: Neural network precision tuning. In: Parker, D., Wolf, V. (eds.) QEST 2019. LNCS, vol. 11785, pp. 129–143. Springer, Cham (2019). https://doi.org/10.1007/978-3-030-30281-8_8

21. Jézéquel, F., Hoseininasab, S., Hilaire, T.: Numerical validation of half precision simulations. In: Rocha, Á., Adeli, H., Dzemyda, G., Moreira, F., Ramalho Correia, A.M. (eds.) WorldCIST 2021. AISC, vol. 1368, pp. 298–307. Springer, Cham (2021). https://doi.org/10.1007/978-3-030-72654-6_29

22. Kotipalli, P.V., Singh, R., Wood, P., Laguna, I., Bagchi, S.: AMPT-GA: automatic mixed precision floating point tuning for GPU applications. In: Proceedings of the ACM International Conference on Supercomputing, Phoenix, Arizona, pp. 160–170. ACM, June 2019

23. Laguna, I., Wood, P.C., Singh, R., Bagchi, S.: GPUMixer: performance-driven floating-point tuning for GPU scientific applications. In: Weiland, M., Juckeland, G., Trinitis, C., Sadayappan, P. (eds.) ISC High Performance 2019. LNCS, vol. 11501, pp. 227–246. Springer, Cham (2019). https://doi.org/10.1007/978-3-030-20656-7_12

24. Lam, M.O., Hollingsworth, J.K., de Supinski, B.R., Legendre, M.P.: Automatically adapting programs for mixed-precision floating-point computation. In: Proceedings of the 27th International ACM Conference on International Conference on Supercomputing, ICS 2013, pp. 369–378. ACM, New York (2013)

25. Lam, M.O., Vanderbruggen, T., Menon, H., Schordan, M.: Tool integration for source-level mixed precision. In: 2019 IEEE/ACM 3rd International Workshop on Software Correctness for HPC Applications (Correctness), pp. 27–35 (2019)

26. Lauter, C., Volkova, A.: A framework for semi-automatic precision and accuracy analysis for fast and rigorous deep learning. arXiv:2002.03869 [cs] (2020)

27. Lin, W., et al.: Robustness verification of classification deep neural networks via linear programming. In: Conference on Computer Vision and Pattern Recognition (2019)

28. Madry, A., Makelov, A., Schmidt, L., Tsipras, D., Vladu, A.: Towards deep learning models resistant to adversarial attacks. In: 6th International Conference on Learning Representations, ICLR (2019). arXiv:1706.06083

29. Mellempudi, N., Srinivasan, S., Das, D., Kaul, B.: Mixed precision training with 8-bit floating point. arXiv:1905.12334 [cs, stat] (2019)

30. Menon, H., et al.: ADAPT: algorithmic differentiation applied to floating-point precision tuning. In: SC18: International Conference for High Performance Computing, Networking, Storage and Analysis, Dallas, TX, USA, pp. 614–626. IEEE, November 2018

31. Na, T., Ko, J.H., Kung, J., Mukhopadhyay, S.: On-chip training of recurrent neural networks with limited numerical precision. In: 2017 International Joint Conference on Neural Networks (IJCNN), Anchorage, AK, USA, pp. 3716–3723. IEEE, May 2017

32. Rakin, A.S. et al.: RA-BNN: constructing robust & accurate binary neural network to simultaneously defend adversarial bit-flip attack and improve accuracy. arXiv:2103.13813 [cs, eess] (2021)
33. Rubio-González, C.et al.: Precimonious: tuning assistant for floating-point precision. In: Proceedings of the International Conference on High Performance Computing, Networking, Storage and Analysis, SC 2013, pp. 27:1–27:12. ACM, New York (2013)
34. Singh, G., Gehr, T., Mirman, M., Püschel, M., Vechev, M.: Fast and effective robustness certification. In: Advances in Neural Information Processing Systems 31: Annual Conference on Neural Information Processing Systems, NeurIPS, pp. 10825–10836 (2018)
35. Su, C., Zhou, S., Feng, L.. Zhang, W.: Towards high performance low bitwidth training for deep neural networks, J. Semicond. **41**, 022404 (2020). https://iopscience.iop.org/article/10.1088/1674-4926/41/2/022404
36. Tjeng, V., Xiao, K., Tedrake, R.: Evaluating robustness of neural networks with mixed integer programming. arXiv:1711.07356 [cs] (2019)
37. Vignes, J.: A stochastic arithmetic for reliable scientific computation. Math. Comput. Simul. **35**, 233–261 (1993)
38. Vignes, J.: Discrete stochastic arithmetic for validating results of numerical software. Numer. Algorithms **37**, 377–390 (2004)
39. Wang, N., Choi, J., Brand, D., Chen, C.-Y., Gopalakrishnan, K.: Training deep neural networks with 8-bit floating point numbers. In: Bengio, S., Wallach, H., Larochelle, H., Grauman, K., Cesa-Bianchi, N., Garnett, R. (eds.) Advances in Neural Information Processing Systems 31, pp. 7686–7695. Curran Associates Inc. (2018). arXiv:1812.08011. http://papers.nips.cc/paper/7994-training-deep-neural-networks-with-8-bit-floating-point-numbers.pdf
40. Zeller, A., Hildebrandt, R.: Simplifying and isolating failure-inducing input. IEEE Trans. Softw. Eng. **28**, 183–200 (2002)
41. Zombori, D.: Verification of artificial neural networks via MIPVerify and SCIP, SCAN (2020)
42. Zombori, D., Bánhelyi, B., Csendes, T., Megyeri, I., Jelasity, M.: Fooling a complete neural network verifier. In: The 9th International Conference on Learning Representations (ICLR) (2021)

MLTL Multi-type (MLTLM): A Logic for Reasoning About Signals of Different Types

Gokul Hariharan[(✉)] [iD], Brian Kempa [iD], Tichakorn Wongpiromsarn [iD],
Phillip H. Jones [iD], and Kristin Y. Rozier [iD]

Iowa State University, Ames, USA
{gokul,bckempa,nok,phjones,kyrozier}@iastate.edu

Abstract. Modern cyber-physical systems (CPS) operate in complex systems of systems that must seamlessly work together to control safety- or mission-critical functions. Capturing specifications in a logic like LTL enables verification and validation of CPS requirements, yet an LTL formula specification can imply unrealistic assumptions, such as that all signals populating the variables in the formula are of type Boolean and agree on a standard time step. To achieve formal verification of CPS systems of systems, we need to write validate-able requirements that reason over (sub-)system signals of different types, such as signals with different timescales, or levels of abstraction, or signals with complex relationships to each other that populate variables in the same formula. Validation includes both transparency for human validation and tractability for automated validation, e.g., since CPS often run on resource-limited embedded systems. Specifications for correctness of numerical algorithms for CPS need to be able to describe global properties with precise representations of local components. Therefore, we introduce Mission-time Linear Temporal Logic Multi-type (MLTLM), a logic building on MLTL, to enable writing clear, formal requirements over finite input signals (e.g., sensor signals, local computations) of different types, cleanly coordinating the temporal logic and signal relationship considerations without significantly increasing the complexity of logical analysis, e.g., model checking, satisfiability, runtime verification (RV). We explore the common scenario of CPS systems of systems operating over different timescales, including a detailed analysis with a publicly-available implementation of MLTLM.

We contribute: (1) the definition and semantics of MLTLM, a lightweight extension of MLTL allowing a single temporal formula over variables of multiple types; (2) the construction and use of an MLTLM fragment for time-granularity, with proof of the language's expressive power; and (3) the design and empirical study of an MLTLM runtime engine suitable for real-time execution on embedded hardware.

1 Introduction

Design and verification of safety-critical systems, such as aircraft, spacecraft, robots, and automated vehicles, requires precise, unambiguous specifications that enable automated reasoning such as model checking, synthesis, requirements debugging, runtime

Artifacts for reproducibility appear at: http://laboratory.temporallogic.org/research/NSV2022.
Funded in part by NSF: CPS Award #2038903, NSF:CAREER Award #1664356, and NASA Cooperative Agreement Grant #80NSSC21M0121.

O. Isac et al. (Eds.): NSV 2022/FoMLAS 2022, LNCS 13466, pp. 187–204, 2022.
https://doi.org/10.1007/978-3-031-21222-2_11

verification (RV), and checking for satisfiability, reachability, realizability, vacuity, and other important properties of system requirements. Modern, cyber-physical systems-of-systems present a unique challenge for specification, and consequently for scalable verification and validation, due to their distributed and hierarchical nature. To seed automated reasoning for CPS systems-of-systems, we need to be able to seamlessly construct global properties combining local phenomena and coordinate requirements for numerical computations like supervision and signal processing over data and variables of different types and sampling frequencies.

Due to the popularity of timelines in operational concepts for CPS systems-of-systems LTL provides an intuitive way to precisely specify system requirements. The relative computational efficiency of automated reasoning (e.g., model checking, satisfiability checking) adds to the appeal of LTL as a specification logic. Since CPS specifications most often need to describe finite missions with referenceable time steps, variations of LTL over finite signals (sometimes also called "traces") emerged with intervals on the temporal operators. Variations on Metric Temporal Logic (MTL) [28], such as Signal Temporal Logic (STL) [16] and Mission-time Linear Temporal Logic (MLTL) [23,29] vary widely in the types of finite bounds they introduce on LTL's temporal operators and the complexity of automated reasoning (e.g., model checking, satisfiability checking) over these logics. MLTL, which adds finite, closed, integer bounds on LTL's temporal operators, has emerged as a popular specification logic for complex CPS systems-of-systems such as the NASA Lunar Gateway Vehicle System Manager [12], and a JAXA autonomous satellite mission [27]; see [24] for a collection of MLTL patterns over a weather balloon, automated air traffic management system, sounding rocket, and satellite. Again, we see the selection of MLTL center on the balance of expressiveness with computational efficiency; MLTL efficiently reduces to LTL [23] and recent work has contributed very efficient, flight-certifiable, encodings of MLTL for runtime verification in resource-limited embedded hardware [20].

However, realistic requirements for CPS systems-of-systems need to combine variables of different types in the same requirement. For example, a requirement specified as an LTL formula may implicitly presume that the input signals populating its atomic propositions share a common notion of a time step. But we struggle to write a single formula to describe a global property about a system where different sub-systems operate at different times, or, more generally, over different types with a non-obvious comparison function. For one example, this problem emerges when we try to specify global safety properties of deep-space-exploring craft. One subsystem of the spacecraft may regulate monthly cycles to wake from hibernation and execute course corrections whereas another subsystem may operate on the nanosecond frequency to make hypersensitive adjustments; it is not obvious how to efficiently reason about these in the same formula. Numerical computations and reasoning on embedded hardware are essential features of CPS, yet they present even more challenges for combining multiple types in a single specification. During long, complex, numerical simulations having a monitor verify statistical patterns in generated data will help detect errors or non-convergence in the early phases, saving computational resources, manual inspection and inefficient postmortem analysis [14,15].

Previous works provide some options for special cases of this problem, with significant complexity drawbacks. These largely center on two philosophies: higher-order logics reasoning over sets of formulas (instead of one formula combining different types), and annotations to deal with multiple time granularities across formula variables, though not necessarily other combinations of different types. Examples of distributed sets of specifications count on locally evaluating sub-system-level synchronous [6] or asynchronous [5,26] signals; this set can coordinate through a global formula evaluated over the local formulas [6]. HyperLTL focuses on specifications over sets of formulas over signals of the same type [9], oppositely from this work where we focus on constructing single formulas that seamlessly reason over signals of different types.

The particular instance of different types in the form of input signals over different time granularities that comprise parts of the same, single temporal logic specification arises frequently in CPS; see [17] for a survey. Most previous works focus on developing well expressible languages to define temporally distributed specifications precisely. Again, this often comes with higher-order reasoning (see for example [18]) and complexity penalties; e.g., [10] introduces the notion of temporal universes and uses a set-theory representation of different timescales to abstract notions of time granularities. Propositional Interval Temporal Logic (PITL) adds chop (";") and project operators to LTL to increase expressivity for time granularities over infinite signals; another variation adds temporal relations like "just before" [11]. First-Order Theory (FOT) enables writing time-granular specifications to account for continuous-in-time events and relate them to discretized-in-time representations [3]. Other methods include using automata to represent time-granularity [21,22] and using spider-diagram representations for time-granular specifications [7], and a two-dimensional metric temporal logic that can be potentially used to represent time granularities [4]. Table 1 collects this related work.

Table 1. Various time-granular specification languages and their syntax elements.

Time-granular logic	Syntax elements	Ref.
PITL	empty, proj, ";", $\Box$, $\Diamond$	[8]
Non standard FOT	$\forall$, $<_e$, $<_w$, $<_1$, $\exists$	[3]
ITL	$<$, m, O, s, f etc.	[1]
Euzenate's extension	6×6 table of operators	[11]
Automata representation	Automata	[21,22]
Spider diagrams	Spider diagrams	[7]
2D MTL internal eternal	L_i, L_e etc.	[4]
Monodic SOL	Layered representation of FOT	[18]

None of the existing solutions enable directly and intuitively specifying linear temporal logic over finite signals containing different types. We need a logic designed for this use, that enables direct specification of common CPS requirements, e.g., for supervision or signal processing, without kludgey syntax that makes correct specifications hard to write, unintuitive constructions that make specifications hard for CPS designers to validate, or introducing complexity blow-ups that make verification techniques like model checking or runtime verification intractable. Therefore, we build upon the popular logic MLTL to create MLTLM, a logic for intuitively and directly expressing

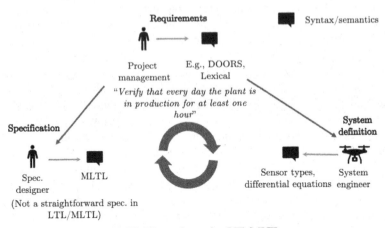

(a) Workflow when using LTL/MLTL

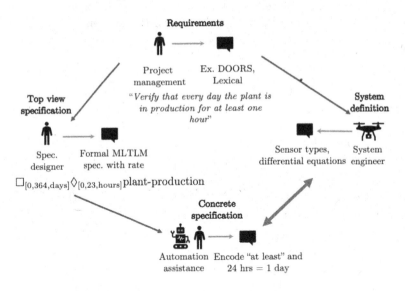

(b) Envisioned workflow using MLTLM

Fig. 1. Iteration workflow for CPS runtime verification of project requirements describing the system and specification in simple lexical language. (a) Traditionally, modifications to the system or specification at any level restarts the cycle. (b) We propose that project management first verify a top view specification in a simple syntax while iteration of the detailed specification is contained to system engineering. The automated assistant will provide hints on suggesting the right projection between types.

bounded temporal logic formulas whose variables may be of different types, including different time granularities. The syntax of MLTLM matches that of MLTL except for the single addition of a signal-type label on each temporal operator to signify the output type of that operator. Figure 1 depicts an example MLTLM specification workflow.

We contribute: (1) a formal definition for the logic MLTLM (Mission-time Linear Temporal Logic Multi-type), including syntax and semantics (Sect. 3); (2) a translation of MLTLM to MLTL with a proof of correctness, enabling use of existing MLTL automated reasoning engine (Sect. 4.1); (3) an open-source implementation of a direct encoding of MLTLM for runtime verification, released as an extension of the flight-certifiable R2U2 engine (Sect. 4.2). We choose R2U2 because it is currently the only runtime verification engine that enables real-time analysis of complex algorithms, such as those for numerical software verification, in real time on embedded hardware [20,30].

Section 2 gives a prelude to the conventional single type temporal logic, MLTL, and gives background on R2U2 – an industry-used runtime verification engine for CPS that we will build upon to monitor MLTLM specifications. Section 3 defines our new logic, MLTLM, providing semantics, examples, properties, and use-cases. Section 4 discusses comparisons to a single signal-type logic, and optimization opportunities for automated reasoning using MLTLM. Finally, Sect. 5 discusses conclusions and scope for future work.

2   Preliminaries

This section formalizes signals and trajectories, overviews MLTL which is extended into MLTLM in Sect. 3, and R2U2 which is adapted to monitor MLTLM in Sect. 4.2.

2.1   Signals and Trajectories

Definition 1 *(Signal). A signal σ over an atomic proposition p is defined as the finite sequence $\sigma = a_0, a_1, \ldots$ where $\sigma[i] = a_i \in \{\text{true}, \text{false}\}$ indicates whether p holds at the discrete time instance i. All signals have a type, written $\sigma^{\mathbb{A}}$ for a signal σ with type $\mathbb{A}$.*

Definition 2 *(Trajectory). A trajectory π over atomic propositions $p_0, \ldots, p_n$ is a set of signals, i.e., $\pi = \{\sigma_0, \sigma_1, \ldots, \sigma_n\}$ where σ_i is a signal over p_i. $\pi_p^{\mathbb{A}}[i]$ refers to the ith value of the signal of type $\mathbb{A}$ over atomic proposition p in π.*

In Sect. 3, we impose that binary logical operators can only operate on signals of the same type. We assume that types represent properties such as frequency that are homogeneous across a type. Related work in linear temporal logic use "traces" or "computations" [2,9], which is typically described as a sequence of sets of atomic propositions. In contrast, we generalize "traces" by allowing member signals to be of different types and call them collectively as a trajectory.

2.2 MLTL

MLTL is a variant of LTL [2] on finite signals with closed temporal bounds [29, 30] on natural numbers.

Definition 3 *(MLTL Syntax [29]). The syntax of an MLTL formula φ over a set of atomic propositions $\mathcal{AP}$ is recursively defined as:*

$$\varphi := \text{true} \mid p \mid \neg\varphi_1 \mid \varphi_1 \wedge \varphi_2 \mid \varphi_1 \mathcal{U}_I \varphi_2$$

where $p \in \mathcal{AP}$, φ_1 and φ_2 are MLTL formulas, $I := [lb, ub]$ is a closed interval bound, such that lb and ub are natural numbers such that $lb \leq ub$.

Abstract Syntax Tree (AST) The AST representation of an MLTL formula has nodes of logical operators and leaves of atomic propositions connected to represent the recursive structure of the expression from Definition 3.

Definition 4 *(MLTL Semantics [29]). The evaluation of an MLTL formula φ on a trajectory π where all signals have uniform type produces a signal σ defined recursively on the signals σ_1 and σ_2 representing the evaluation of its child subformula(s) φ_1 and φ_2 respectively.*

$$\sigma[i] := \begin{cases} \pi_p[i] & \text{if } \varphi = p \\ \neg\sigma_1[i] & \text{if } \varphi = \neg\varphi_1 \\ \sigma_1[i] \wedge \sigma_2[i] & \text{if } \varphi = \varphi_1 \wedge \varphi_2 \\ \text{true } \textit{iff } |\sigma_1|, |\sigma_2| > (i + ub) \text{ and} & \\ \quad \exists j \in [i+lb, i+ub] \text{ such that } \sigma_2[j] = \text{true} & \text{if } \varphi = \varphi_1 \mathcal{U}_{[lb,ub]} \varphi_2 \\ \quad \text{and } \forall k < j \text{ where } k \in [i+lb, i+ub], \sigma_1[k] = \text{true} & \end{cases}$$

Other common operators are defined via equivalences, i. e., false $\Leftrightarrow \neg$true, future $\Diamond_I\varphi \Leftrightarrow \text{true}\,\mathcal{U}_I\,\varphi$, globally $\Box_I\varphi \Leftrightarrow \neg(\Diamond_I\neg\varphi)$, and next $\bigcirc\varphi \Leftrightarrow \Box_{[1,1]}\varphi$.

2.3 R2U2

The *Realizable, Responsive, Unobtrusive Unit*[1] (R2U2) is an MLTL based RV engine for flight mission systems [29] used in robotics [20], NASA drone aircraft [19,31], and is being evaluated for use on the Lunar Gateway space station [12]. R2U2 is *Realizable*: implemented on real hardware, *Responsive*: reports specification violation immediately, and *Unobtrusive*: uses existing data sources instead of modifying the system to add instrumentation. R2U2 features specification reconfiguration and real-time performance with guaranteed memory bounds to better support the needs of flight systems. We have developed our MLTLM verification engine upon R2U2, which is an open-source RV engine with well-documented industrial use to provide users with a seamless move to multi-type logic. The R2U2 based MLTLM verification engine we develop upholds all existing guarantees of R2U2.

---

[1] r2u2.temporallogic.org.

3   Mission-Time Temporal Logic Multi-type (MLTLM)

We develop the foundations of MLTLM in this section. MLTLM is a lightweight extension to MLTL that enables temporal reasoning over system trajectories composed of signals of different types.

Definition 5 *(MLTLM Syntax). The syntax of an MLTLM formula φ over a set of atomic propositions $\mathcal{AP}$ is recursively defined as:*

$$\varphi := \mathrm{true} \mid p \mid \neg\varphi \mid \varphi_1 \wedge \varphi_2 \mid \varphi_1 \mathcal{U}_J \varphi_2$$

where $p \in \mathcal{AP}$, φ_1 and φ_2 are MLTLM formulas, and $J := [lb, ub, \mathbb{A}]$ is a finite interval bound such that lb and ub are natural numbers, $lb \leq ub < \infty$, and $\mathbb{A}$ is a label indicating the signal type over which an MLTLM temporal operator evaluates.

Notably, MLTLM syntax is MLTL syntax with signal types associated with temporal operators.

Definition 6 *(MLTLM Semantics). The evaluation of an MLTLM formula φ on a trajectory π produces a signal σ of type $\mathbb{A}$ defined recursively on the signals σ_1 and σ_2 representing the evaluation of its child subformula(s) φ_1 and φ_2 respectively.*

$$\sigma^{\mathbb{A}}[i] := \begin{cases} \pi_p^{\mathbb{A}}[i] & \text{if } \varphi = p \\ \neg\sigma_1^{\mathbb{A}}[i] & \text{if } \varphi = \neg\varphi_1 \\ \sigma_1^{\mathbb{A}}[i] \wedge \sigma_2^{\mathbb{A}}[i] & \text{if } \varphi = \varphi_1 \wedge \varphi_2 \\ \sigma_1^{\mathbb{A}}[i..] \, \mathcal{U}_{[lb,ub]} \sigma_2^{\mathbb{A}}[i..] & \text{if } \varphi = \varphi_1 \mathcal{U}_{[lb,ub,\mathbb{A}]} \varphi_2 \end{cases}$$

where $\sigma[i..]$ is the subsequence of signal σ starting from discrete point i and all operators are evaluated according to the rules of Definition 4.

Note that when evaluating the fourth case in Definition 6, the signal types produced by subformulas φ_1 and φ_2 must be projected into signals of the type associated with the temporal operator. Additional common operators like implication, disjunction, and globally are constructed by standard equivalence relations as in MLTL, with all derived temporal operators inheriting the type specifier on their interval bounds. If the relationship between types can be expressed as a function that converts the type of signals, that function is called a projection.

Definition 7 **(Projection).** *The projection function $T_{\mathbb{A}}^{\mathbb{B}}(\sigma^{\mathbb{A}})$ takes the signal σ of type $\mathbb{A}$ and returns a new signal of type $\mathbb{B}$.*

We will examine several projection functions, however writing MLTLM formulas requires only assurance their existence, not their definition; this provides a separation of concerns we leverage to ease specification writing and linearize verification workflow. For example, consider a formula φ specifying that φ_1 should hold every hour for 10 h, and φ_2 should hold every second for 100 s. In MLTLM φ could be written as $\Box_{[0,9,\mathrm{hour}]}\varphi_1 \wedge \Box_{[0,99,\mathrm{second}]}\varphi_2$. In MLTL, φ would need to be written assuming a monitor rate, say seconds, then specifier would write $\Box_{[0,0]}\varphi_1 \wedge \Box_{[3600,3600]}\varphi_1$

$\wedge \ \Box_{[7200,7200]}\varphi_1 \ \wedge \cdots$ and $\Box_{[0,99]}\varphi_2$. The formula is longer and embeds the relation between hours and seconds. If the specification must be evaluated at a monitor rate of minutes instead, the canonical encoding must be updated by the specification author as discussed in more detail in Sect. 1 (Fig. 1). In contrast, in MLTLM, the top view specification remains the same even in the face of implementation details like evaluation rate.

3.1 Equivalent MLTLM Formula for Every MLTL Formula

For a formula naming at most one type, all properties that hold in MLTL hold in MLTLM, i.e., $\Diamond_{[lb,ub,\mathbb{A}]}\varphi \Leftrightarrow \text{true } \mathcal{U}_{[lb,ub,\mathbb{A}]} \ \varphi$, $\Box_{[lb,ub,\mathbb{A}]}\varphi \Leftrightarrow \neg(\Diamond_{[lb,ub,\mathbb{A}]}\neg\varphi)$ and so on. The following claim expresses that formulas expressible in MLTL form a subset of formulas expressible in MLTLM. The claim attests that there is no loss in using MLTLM compared to MLTL. The transformation is simple, and the formula is, at worst, the same length, though potentially much shorter in MLTLM, as demonstrated in Sect. 4.3.

Claim. An equivalent MLTLM formula of the same length exists for every MLTL formula, and this translation is possible in constant time.

Proof. We can represent any MLTL formula as an MLTLM formula by appending a signal type to the interval bound of every temporal operator. This follows from the definition of MLTLM. The formula length, being the total number of operators plus atomic propositions, is not affected by appending a type name to the temporal operators. Hence the resultant MLTLM formula is of the same length as the MLTL formula.

3.2 Evaluation of MLTLM Formula

Evaluation of an MLTLM formula on a trajectory requires signals for all atomic propositions. Evaluating an MLTLM formula naming at most one type over a trajectory is equivalent to evaluating MLTL formulas over a trajectory containing only the required signals.

With projection, a new signal of a different type can be derived from an existing signal in the trajectory. For example, the return of a high-rate sensor can be down-sampled to match the type of low rate sensor. This "derived signal" evaluation is where all signals are first projected to a common type before evaluation. Using signals, types, and projection, we can evaluate a formula with mixed types by considering each subformula to represent the signal of its own evaluation and projecting where necessary as explained further in the next section.

Critically, operator semantics are defined for any type, but only when the input(s) and output types match. The inputs to the temporal logic operator must be projected to the written type in the operator's bound if needed. Fundamentally, MLTLM formulas represent a directed graph of data flow between domains of MLTL connected by projections.

Tutorial Example Application of the MLTLM Semantics (Definition 6). To help clarify how the semantics in Definition 6 are applied, we consider the formula

$\square_{[1,2,\mathbb{B}]}(\square_{[2,4,\mathbb{A}]}p)$. The global ($\square$) operator is a common unary temporal operator derived from the definition of $\mathcal{U}$ by the equivalence relation $\square_{[lb,ub,\mathbb{A}]}\varphi \Leftrightarrow \neg(\text{true } \mathcal{U}_{[lb,ub,\mathbb{A}]} \neg\varphi)$. This is the same as adding the following case to the MLTLM semantics:

$$\sigma[i] := \text{true iff } \sigma_1^{\mathbb{A}}[j] = \text{true } \forall j \in [i+lb, i+ub], \qquad \text{if } \varphi = \square_{[lb,ub,\mathbb{A}]}\varphi_1.$$

Applying Definition 6, the evaluation of formula $\square_{[1,2,\mathbb{B}]}(\square_{[2,4,\mathbb{A}]}p)$ depends on the type of any known signals for p and the desired output type. Let us consider generating a signal of type $\mathbb{B}$ from the above formula, and that $\pi_p^{\mathbb{A}}$ is known for p. In Fig. 2a, the known signal for p, $\sigma_1^{\mathbb{A}}$, is input to $\square_{[2,4,\mathbb{A}]}$ whose satisfaction signal, $\sigma_2^{\mathbb{A}}$, is input to $\square_{[1,2,\mathbb{B}]}$, finally generating $\sigma_1^{\mathbb{B}}$ which meets the required output of type $\mathbb{B}$.

Now let us consider another case with the same formula where we need an output signal of type $\mathbb{C}$, and know $\pi_p^{\mathbb{B}}$. In Fig. 2b, evaluating the subformula $\square_{[2,4,\mathbb{A}]}p$ requires a signal for p in type $\mathbb{A}$ per the semantics, but we only know p in type $\mathbb{B}$. This implies a projection $T_{\mathbb{B}}^{\mathbb{A}}(\sigma_1^{\mathbb{B}}) = \sigma_1^{\mathbb{A}}$ before the result is input to $\square_{[2,4,\mathbb{A}]}$, generating $\sigma_2^{\mathbb{A}}$. Another type incompatibility arises between $\sigma_2^{\mathbb{A}}$ and $\square_{[1,2,\mathbb{B}]}$, so it is again (implicitly) projected to a type $\mathbb{B}$ through $T_{\mathbb{A}}^{\mathbb{B}}(\sigma_2^{\mathbb{A}}) = \sigma_3^{\mathbb{B}}$. Since the desired output type is $\mathbb{C}$, there is one last projection $T_{\mathbb{B}}^{\mathbb{C}}(\sigma_3^{\mathbb{B}}) = \sigma_1^{\mathbb{C}}$.

3.3   Examples of Projections

Earlier in Sect. 3 we defined an abstract projection (Definition 7). This section will consider a couple of useful projections and discuss some example specifications.

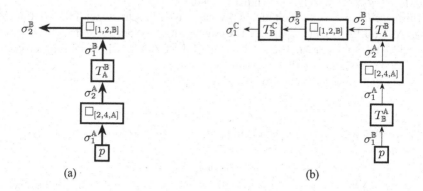

(a)   (b)

Fig. 2. Illustration of two possible evaluations of a given formula $\square_{[1,2,\mathbb{B}]}(\square_{[2,4,\mathbb{A}]}p)$.

Definition 8. *(Modulo-Reduction Function). The function $f_s : \sigma^{\mathcal{A}} \to \sigma^{\mathcal{B}}$ implements the projection $T_{\mathbb{A}}^{\mathbb{B}}(\sigma^{\mathcal{A}})$ by modulo-reduction with positive integer stride s when:*

$$f_s(\sigma^{\mathcal{A}}) = \sigma^{\mathcal{B}} \text{ such that } \sigma^{\mathcal{B}}[i] = \sigma^{\mathcal{A}}[i \cdot s] \tag{1}$$

The modulo-reduction function outputs every sth value from the input signal.

Definition 9. *(Majority-Reduction Function). The function $g_s : \sigma^{\mathbb{A}} \to \sigma^{\mathbb{B}}$, implements the projection $T_{\mathbb{A}}^{\mathbb{B}}(\sigma^{\mathbb{A}})$ by majority-reduction with positive integer stride s when:*

$g_s(\sigma^{\mathbb{A}}) = \sigma^{\mathbb{B}}$ *such that*

$$\sigma^{\mathbb{B}}[i] = \begin{cases} \text{true} & \text{if } \mathcal{N}_0(\{j \in [i \cdot s, (i+1) \cdot s] : (\sigma^{\mathbb{A}}[j] = \text{true})\}) \geq \lfloor s/2 \rfloor \\ \text{false} & \text{otherwise} \end{cases} \quad (2)$$

where $\mathcal{N}_0(\cdot)$ is the set cardinality.

The majority-reduction function outputs the majority value of every s values of the input signal.

3.4   Example Specifications Across Timescales

We consider a few example specifications taken from literature on time-granularities [17,25], and modify or extend them to the context of RV.

1. "Verify that John is present for 8 h at a stretch each day for the next 6 days."
   This specification can be represented in MLTLM as:

   $$\Box_{[0,5,\text{day}]} (\Diamond_{[0,16,\text{hour}]} \Box_{[0,7,\text{hour}]} \text{john-present}) \quad (3)$$

   The specification says that eventually, from the 0th to the 16th hour, there exists an hour such that John is present from the 0th to the 7th hour. The eventually operator has a time going from 0 to 16, and the global operator from 0 to 7, and the total time adds to 0 to 23 h, which is a 24 h period (a day).
   This specification is verified on a daily basis, based on the type of the root node of the AST for Eq. (3), the $\Box_{[0,5,\text{day}]}$. The day type must be projected from the hour type used by the subformula. The satisfaction of the formula depends on the projection used to go from the hourly type to the daily type.
2. "Verify that for at least one day in a year the plant works every hour"

   $$\Diamond_{[0,364,\text{day}]} \Box_{[0,23,\text{hour}]} \text{plant-works}$$

3. "Verify that every day the plant is in production for some hours"

   $$\Box_{[0,364,\text{day}]} \Diamond_{[0,23,\text{hour}]} \text{plant-production}$$

4. "Verify that the plant is monitored by the remote system every minute of every hour for the next 24 h"

   $$\Box_{[0,23,\text{hour}]} \Box_{[0,59,\text{minute}]} \text{system-monitored}$$

5. "On all days of the year, the plant works for at least 12 h"
   We represent this in MLTLM using the majority-reduction function (Definition 9), with $\mathbb{A} \equiv$ hour and $\mathbb{B} \equiv$ day as

   $$\Box_{[0,364,\text{day}]} \Box_{[0,0,\text{hour}]} \text{plant-works}$$

6. "Verify that the system deviates at most for a minute every hour for the next 24 h."
   We can represent this in MLTLM by modifying the cardinality relation in Eq. (2) to
   "> 1" and using the resultant function with $\mathbb{A} \equiv$ hour and $\mathbb{B} \equiv$ day as the projection,

$$\square_{[0,23,\text{hour}]}\square_{[0,0,\text{minute}]}\text{system-deviates}$$

4   Equisatisfiable Formula in MLTL and an Implementation of an MLTLM Monitor with the Modulo-Reduction Projection

The previous section introduced MLTLM and demonstrated how it could simplify the workflow and specifications across timescales. We now illustrate space and time optimization possibilities by implementing an MLTLM RV engine. The generic syntax and semantics of MLTLM separates the specification from the signal type, i.e., the specification remains the same irrespective of the signal type. It is apparent from the semantics (Definition 6) that the output signal type is determined only in the fourth case with the temporal operator. For example, the formula $p \wedge q$ represents multiple output signal types depending on the trajectory types used for p and q, whereas the formula $\square_{[0,0,\mathbb{A}]}(p \wedge q)$ has a single output type $\mathbb{A}$ irrespective of the trajectory types used for p and q. An implementation needs a single output type, and hence we consider a subset of MLTLM formulas that have a temporal operator at the root of the AST, and assume that the type on the root temporal operator is the desired output type.

Furthermore, to make the evaluation of an MLTLM formula complete, two more ingredients are essential, (a) the placement of projections in the AST of an MLTLM formula and (b) defined projections between type signals. Consider the MTLTM formula, $\square_{[0,0,\mathbb{A}]}(p \wedge q)$. Let us assume that only a signal of type $\mathbb{B}$ is available from p and a signal of type $\mathbb{C}$ is available from q, as denoted in Fig. 3a. From the semantics Definition 6, it is clear that a conjunction is allowed only between signals of the same type, which implies that there are implicit projections to match signal types in the conjunction as shown in Fig. 3b.

We have two (out of many) options here to match types, (a) to project to a common signal type $\mathbb{D}$ at the conjunction, and then to a type $\mathbb{A}$ to match type in $\square_{[0,0,\mathbb{A}]}$ (Fig. 3b), and (b) place a projection to type $\mathbb{A}$ at the conjunction, then a second projection is not needed to match types in $\square_{[0,0,\mathbb{A}]}$ (Fig. 3c). While option (a) is of interest in the broader scope of applications with MLTLM like signal processing, option (b) is the situation with the minimal number of projections. *The generalization for this minimal projection placement is to impose that signals are projected to the type of the closest ancestor node with a type.* All nodes in the unique path connecting a node to the root of the AST are ancestor nodes of the node (the node inclusive). In this example, the closest ancestor of the conjunction is $\square_{[0,0,\mathbb{A}]}$ whose type is $\mathbb{A}$. We further assume that all such projections exist to evaluate a formula.

We consider only the modulo-reduction function (Definition 8) as it is not possible to cover all scenarios in this paper. We develop a theory to derive equisatisfiable MLTL formula for an MLTLM formula with a class of logical projections and then develop a translator based on it with the modulo-reduction projection (Definition 8). We then compare the memory and time needed to evaluate formulas using MLTLM and MLTL.

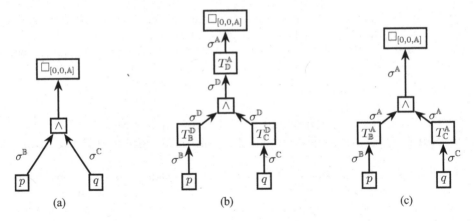

Fig. 3. The evaluation of an MLTLM formula depends on the placement of projections to match types in binary operators.

In summary, we find that MLTLM reflects on profound savings in memory compared to its closest single-type logic. The results presented herein are only preliminary observations of optimization possibilities using MLTLM.

4.1 The Translator

Theorem 1 *(Expressive Equivalence of MLTL and MLTLM with Logical Projections).* *Let F be a projection expressible in MLTL, then F is a logical projection. Let $\mathbb{A}$ be a type and ψ be an MLTL formula that outputs signals of type $\mathbb{A}$. For every MLTLM formula φ such that for every type t in φ there exists a chain of logical projections from t to $\mathbb{A}$, the signal generated by φ is equivalent to another signal generated by ψ.*

Proof Sketch. The full proof is available in supplementary material posted online[2]. We give an example sketch over two signal types $\mathbb{B}$ and $\mathbb{C}$ related by a logical projection $F(\sigma^{\mathbb{B}}) = \sigma^{\mathbb{C}}$. We use the semantics of MLTLM (Definition 6) to prove by induction on the structure of the formula that any MLTLM formula that has temporal operators with both types $\mathbb{B}$ and $\mathbb{C}$ can be reduced to an equisatisfiable formula all of whose temporal operators are of type $\mathbb{B}$. An MLTLM formula of a single type can be reduced to an MLTL formula by merely removing the type from the formula.

We complete the proof by assuming that a formula of the form $\varphi = \varphi_1 \mathcal{U}_{[lb,ub,\mathbb{C}]} \varphi_2$ can be equivalently expressed with type $\mathbb{B}$ in its AST root using the logical projection. For example, we can show that the modulo-reduction projection (Definition 8) can be equivalently expressed as an MLTL formula without a projection to type $\mathbb{C}$ using the

---

[2] http://laboratory.temporallogic.org/research/NSV2022.

function $p(\varphi)$ where

$$p(\varphi) = \bigcirc_{\mathbb{B}}^{lb}(\varphi_2$$
$$\vee \; (\varphi_1 \wedge \bigcirc_{\mathbb{B}}^{s}\varphi_2)$$
$$\vee \; (\varphi_1 \wedge \bigcirc_{\mathbb{B}}^{s}\varphi_1 \wedge \bigcirc_{\mathbb{B}}^{2s}\varphi_2)$$
$$\vee \; (\varphi_1 \wedge \bigcirc_{\mathbb{B}}^{s}\varphi_1 \wedge \bigcirc_{\mathbb{B}}^{2s}\varphi_1 \wedge \bigcirc_{\mathbb{B}}^{3s}\varphi_2) \qquad\qquad (4)$$
$$\vdots$$
$$\vee \; (\varphi_1 \wedge \bigcirc_{\mathbb{B}}^{s}\varphi_1 \wedge \bigcirc_{\mathbb{B}}^{2s}\varphi_1 \wedge \bigcirc_{\mathbb{B}}^{3s}\varphi_1 \cdots \wedge \bigcirc_{\mathbb{B}}^{(m-1)s}\varphi_1 \wedge \bigcirc_{\mathbb{B}}^{ms}\varphi_2) \;),$$

where $\bigcirc_{\mathbb{B}} = \square_{[1,1,\mathbb{B}]}$ is the next operator (and hence, $\bigcirc_{\mathbb{B}}^{s} = \square_{[s,s,\mathbb{B}]}$), and $m = \lfloor (ub - lb)/s \rfloor$. We then extend this to all cases in the semantics (Definition 6). Thus, any MLTLM formula φ with mixed signal types $\mathbb{B}$ and $\mathbb{C}$ has an equisatisfiable formula $q(\varphi)$, where the entire formula has a single type, $\mathbb{B}$, defined recursively by

$$q(\varphi) := \begin{cases} \varphi, & \text{if } \varphi \text{ has only one type, } \mathbb{B} \text{ in the entire formula,} \\ p(\, q(\varphi_1)\mathcal{U}_{[lb,ub,\mathbb{C}]}q(\varphi_2)\,), & \text{if } \varphi = \varphi_1\mathcal{U}_{[lb,ub,\mathbb{C}]}\varphi_2, \\ \neg q(\varphi_1), & \text{if } \varphi = \neg\varphi_1, \\ q(\varphi_1) \wedge q(\varphi_2), & \text{if } \varphi = \varphi_1 \wedge \varphi_2. \end{cases}$$

$$(5)$$

It is straightforward to extend this analysis to multiple types that have transitional chain of connected projections. We showed that we could derive an equisatisfiable formula verifiable in the image type for any MLTLM formula when using the logical projections. We implement a translator from MLTLM to MLTL using the theory discussed.

We developed three translators from MLTLM to MLTL based on the recursive formula Eq. 5. The three translators are based on succinct and expanded versions of Eq. 4 the most succinct (to our best capability) being translator 3, and the most expanded being translator 1. The translator's details and proof of correctness will be reported elsewhere in the interest of space. We confirm, however, that verdicts from the three translators on a well-established MLTL engine (R2U2 [20]) and its extended MLTLM monitor developed by us (discussed in Sect. 4.2) produce consistent outputs for the same inputs with more than 70 randomly generated formulas.

4.2   An Efficient MLTLM Engine

We implement an RV engine for specifications in MLTLM on top of R2U2 (see Sect. 2, and [20]). Certain notes on how specifications are written out for verification using R2U2 are relegated to supplementary material online[3]. We skip the details of the implementation in the interest of space and report it elsewhere. We summarize the implementation briefly.

As we mentioned many times in this article, if a sensor data is only of interest every hour, then the second-to-second information can be dropped out; and the modulo-reduction function (Definition 8) does this operation. The MLTLM engine has added

---

[3] http://laboratory.temporallogic.org/research/NSV2022.

projection operators (see Definition 7) at appropriate places according to the semantics of MLTLM (Definition 6) using the closest ancestor type projection discussion in Sect. 4 (Fig. 3). The modulo-reduction projection operator drops the appropriate signal values not needed in evaluating a formula and reports the output signal type corresponding to the type in the root of the AST.

4.3   Optimization Results

In Sect. 3.1 we showed that every MLTL formula could be expressed in MLTLM in the same length. This section analyzes how long are the intuitive translations to MLTL compared to MLTLM. We do not claim rigorous proof on the shortest possible formula length but rather compare the most intuitive and succinct translations. We note that the translations contain expressions of the form (see Eq. 4),

$$\varphi_1 \wedge \bigcirc_{\mathbb{B}}^{s} \varphi_1 \wedge \bigcirc_{\mathbb{B}}^{2s} \varphi_1 \wedge \bigcirc_{\mathbb{B}}^{3s} \varphi_2,$$

which to the best of our knowledge cannot be made any shorter in LTL and MLTL [32].

We randomly draw MLTLM formulas using the procedure in [13] and plot the length of the shortest intuitive MLTL translations. The randomly drawn formulas are parametrized by the probability of drawing a temporal operator (P), the maximum difference between the lower and upper bounds (M), and the maximum signal length (T). We will fix $M = T = 6$ in our study here. Furthermore, the memory and time also depends on stride, s of the modulo-reduction function (see Eq. (1)). In real systems specifications may reason over say, seconds, minutes, hours and days, which correspond to $s = 60$, and 24. However, as we mentioned previously, we are reporting preliminary observations on optimization possibilities, and we use four signal types, which we will call $\mathbb{A}$, $\mathbb{B}$, $\mathbb{C}$ and $\mathbb{D}$, where (see Eq. (1) for $f_s(\sigma)$), with

$$f_2(\sigma^{\mathbb{A}}) = \sigma^{\mathbb{B}}, \quad f_3(\sigma^{\mathbb{B}}) = \sigma^{\mathbb{C}}, \quad f_4(\sigma^{\mathbb{C}}) = \sigma^{\mathbb{D}},$$

with stride lengths $s = 2, 3, 4$. Note that the memory savings will be much larger with a larger stride like $s = 60$ (e.g., from second to minute).

Figure 4a shows the cumulative formula length with randomly drawn formulas. At $P = 0.5$, the three translators produce MLTL formulas of nearly the same length (the dotted, dashed, and dashed and dotted lines). However, Translator 3 performs slightly better with shorter formula lengths. In contrast, the formula lengths of the MLTLM formulas are substantially smaller. Hence, there is no loss in using MLTLM in comparison to MLTL (see Sect. 3), but using MLTLM may result in much smaller and more intuitive formulas depending on the projection function.

Figure 4b shows the mean formula length (averaged over 60 random formulas) by varying the probability of the temporal operator. $P = 0$ corresponds to no temporal operators, and in that case, the translators and the MLTLM formula perform nearly equally well. This is expected – if the formula contains mere propositional logic, the formula should be independent of the temporal specification language. However, on close observation, the MLTLM formula at $P = 0$ is slightly longer. This is because in MLTL there is only one signal type, hence there is no need for a output signal type specifier, whereas in MLTLM, a proposition (say, $p \wedge q$, $p, q \in \mathcal{AP}$) represents a family of

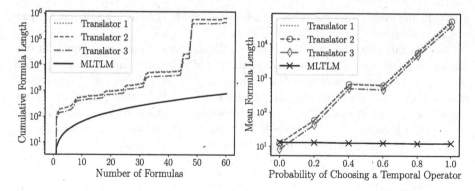

Fig. 4. (a) Cumulative formula length with the number of randomly drawn formulas with $P = 0.5$, and (b) mean formula length against the probability of choosing a temporal operator.

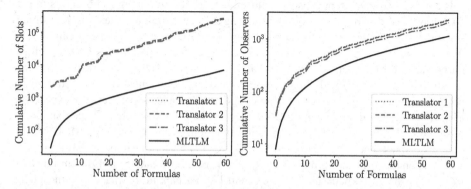

Fig. 5. Cumulative memory (left) and time (right) needed to verify random MLTLM formulas from the benchmark set in Sect. 4 using an equivalent MLTL formula (translator) compared to an MLTLM engine with a Modulo-Reduction projection operator in R2U2.

outputs of different types. We always use a temporal operator at the start of any formula (as in $\square_{[0,0,\mathbb{B}]}(p \wedge q)$ in the place of $p \wedge q$), and this adds to excess length of an MLTLM formula compared to an MLTL formula with propositions. However, propositions like $p \wedge q$ are valid MLTLM formulas, but verification of the formula needs an output-type identifier.

On increasing the probability of choosing a temporal operator, the equisatisfiable formulas in MLTL become significantly longer owing to the expansion to the base type as discussed in Sect. 4.1. Figure 5 shows the estimated resource and time requirements on hardware. The memory to evaluate a formula is statically assigned in R2U2 [20] as dynamic memory is often not permitted in flight software. Hence, we compare the amount of static memory that needs to be assigned for equisatisfiable formulas in MLTLM and (translated) MLTL (Fig. 5a). Similarly, the time taken for formula evaluation is directly proportional to the number of nodes created in the AST. We call the nodes in the AST as observers (as seen in the Y axis labels of Fig. 5b). We see that equisatisfiable formula require much lesser memory in MLTLM than MLTL (Fig. 5a).

Similarly, the evaluation time is also much faster for MLTLM as it needs much lesser observers (Fig. 5b).

We end this section with a few remarks. We considered random formulas in this section, and they may not be true representatives of real specifications that may have different results on memory and time savings (Fig. 5). Nonetheless, the results show that there is great opportunity to have short intuitive formulas that encode timescales directly in the formula to simplify the workflow (Fig. 1), and in addition, an optimally configured RV engine for MLTLM is likely to have profound memory savings making it more suitable for resource constrained hardware.

5   Conclusion

Writing specifications naturally needs reasoning across multiple signal types, be it signals coming from different sensors at different rates, or belonging to observers in parallel universes (distributed systems), or having a mix of continuous and discrete signals (hybrid systems). We developed a multi-type logic to express such specifications, and then explored an application to time granularities. As discussed, this serves multiple purposes: 1) for the user, specifications are easy to write, 2) the theoretical satisfaction in different types is defined unambiguously, and 3) implementations can better utilize resources when compared with a single signal-type logic. Moreover, we expect that MLTLM will simplify the workflow by keeping the syntax simple and accessible, and postponing the nuances into the projection function. More importantly, MLTLM separates the specification from signal type. For example, let us suppose that a pressure sensor is changed in the Lunar Gateway, and it generates data in a different rate than the old sensor, or perhaps in a different unit like Pascals in the place of atmospheric pressure. Specifications for a single type logic would have to be changed to account for the signal type. MLTLM side-steps this process: The signal type will not affect the specification in any manner. In the future, we plan to have an automated assistant, that will allow a user to choose different projections to use for different contexts in specifications, (like "at least", "at most", "only once" etc.), and will also inform the user about the amount of memory he will need to dedicate/save on the hardware (the memory needed may vary based on the type of projection). This will allow the industrial verification community to seamlessly move to a time-granular logic. We will also consider human authored MLTLM specifications on real systems to get a better perspective on optimization opportunities. Lastly, the MLTLM monitor built upon R2U2 was validated across a regression suite of specifications and trajectories, but the current implementation can be improved to have tighter bounds on memory usage, which needs further investigation.

References

1. Allen, J.F., Hayes, P.J.: A common-sense theory of time. In: Proceedings of the 9th International Joint Conference on Artificial Intelligence - Volume 1, IJCAI 1985, pp. 528–531. Morgan Kaufmann Publishers Inc., San Francisco (1985)
2. Baier, C., Katoen, J.P.: Principles of Model Checking. MIT Press, Cambridge (2008)

3. Balbiani, P.: Time representation and temporal reasoning from the perspective of non-standard analysis. In: Proceedings of the Eleventh International Conference on Principles of Knowledge Representation and Reasoning, KR 2008, pp. 695–704. AAAI Press (2008)
4. Baratella, S., Masini, A.: A two-dimensional metric temporal logic. Math. Log. Q. **66**(1), 7–19 (2020). https://doi.org/10.1002/malq.201700036
5. Bataineh, O., Rosenblum, D.S., Reynolds, M.: Efficient decentralized LTL monitoring framework using tableau technique. ACM Trans. Embed. Comput. Syst. **18**(5s), 1–21 (2019)
6. Bauer, A., Falcone, Y.: Decentralised LTL monitoring. In: Giannakopoulou, D., Méry, D. (eds.) FM 2012. LNCS, vol. 7436, pp. 85–100. Springer, Heidelberg (2012). https://doi.org/10.1007/978-3-642-32759-9_10
7. Bottoni, P., Fish, A.: Policy specifications with timed spider diagrams. In: 2011 IEEE Symposium on Visual Languages and Human-Centric Computing (VL/HCC), pp. 95–98 (2011). https://doi.org/10.1109/VLHCC.2011.6070385
8. Bowman, H., Thompson, S.: A decision procedure and complete axiomatization of finite interval temporal logic with projection. J. Log. Comput. **13**(2), 195–239 (2003). https://doi.org/10.1093/logcom/13.2.195
9. Clarkson, M.R., Finkbeiner, B., Koleini, M., Micinski, K.K., Rabe, M.N., Sánchez, C.: Temporal logics for hyperproperties. In: Abadi, M., Kremer, S. (eds.) POST 2014. LNCS, vol. 8414, pp. 265–284. Springer, Heidelberg (2014). https://doi.org/10.1007/978-3-642-54792-8_15
10. Clifford, J., Rao, A.: A simple, general structure for temporal domains (1986)
11. Cohen-Solal, Q., Bouzid, M., Niveau, A.: An algebra of granular temporal relations for qualitative reasoning. In: Proceedings of the 24th International Conference on Artificial Intelligence, IJCAI 2015, pp. 2869–2875. AAAI Press (2015)
12. Dabney, J.B., Badger, J.M., Rajagopal, P.: Adding a verification view for an autonomous real-time system architecture. In: AIAA Scitech 2021 Forum, p. 0566 (2021)
13. Daniele, M., Giunchiglia, F., Vardi, M.Y.: Improved automata generation for linear temporal logic. In: Halbwachs, N., Peled, D. (eds.) CAV 1999. LNCS, vol. 1633, pp. 249–260. Springer, Heidelberg (1999). https://doi.org/10.1007/3-540-48683-6_23
14. Dinh, M.N., Abramson, D., Jin, C.: Runtime verification of scientific codes using statistics. Procedia Comput. Sci. **80**, 1473–1484 (2016). https://doi.org/10.1016/j.procs.2016.05.468. International Conference on Computational Science 2016, ICCS 2016, 6–8 June 2016, San Diego, California, USA
15. Dinh, M.N., Trung Vo, C., Abramson, D.: Tracking scientific simulation using online time-series modelling. In: 2020 20th IEEE/ACM International Symposium on Cluster, Cloud and Internet Computing (CCGRID), pp. 202–211, May 2020. https://doi.org/10.1109/CCGrid49817.2020.00-73
16. Donzé, A.: On signal temporal logic. In: Legay, A., Bensalem, S. (eds.) RV 2013. LNCS, vol. 8174, pp. 382–383. Springer, Heidelberg (2013). https://doi.org/10.1007/978-3-642-40787-1_27
17. Euzenat, J., Montanari, A.: Time granularity. In: Handbook of Temporal Reasoning in Artificial Intelligence, January 2005
18. Franceschet, M., Montanari, A., Peron, A., Sciavicco, G.: Definability and decidability of binary predicates for time granularity. J. Appl. Log. **4**(2), 168–191 (2006). https://doi.org/10.1016/j.jal.2005.06.004
19. Geist, J., Rozier, K.Y., Schumann, J.: Runtime observer pairs and bayesian network reasoners on-board fpgas: flight-certifiable system health management for embedded systems. In: Bonakdarpour, B., Smolka, S.A. (eds.) RV 2014. LNCS, vol. 8734, pp. 215–230. Springer, Cham (2014). https://doi.org/10.1007/978-3-319-11164-3_18

20. Kempa, B., Zhang, P., Jones, P.H., Zambreno, J., Rozier, K.Y.: Embedding online runtime verification for fault disambiguation on Robonaut2. In: Bertrand, N., Jansen, N. (eds.) FORMATS 2020. LNCS, vol. 12288, pp. 196–214. Springer, Cham (2020). https://doi.org/10.1007/978-3-030-57628-8_12

21. Lago, U.D., Montanari, A., Puppis, G.: Compact and tractable automaton-based representations of time granularities. Theor. Comput. Sci. **373**(1), 115–141 (2007). https://doi.org/10.1016/j.tcs.2006.12.014

22. Lago, U.D., Montanari, A., Puppis, G.: On the equivalence of automaton-based representations of time granularities. In: 14th International Symposium on Temporal Representation and Reasoning (TIME 2007), pp. 82–93 (2007). https://doi.org/10.1109/TIME.2007.56

23. Li, J., Vardi, M.Y., Rozier, K.Y.: Satisfiability checking for mission-time LTL. In: Dillig, I., Tasiran, S. (eds.) CAV 2019. LNCS, vol. 11562, pp. 3–22. Springer, Cham (2019). https://doi.org/10.1007/978-3-030-25543-5_1

24. Luppen, Z., et al.: Elucidation and analysis of specification patterns in aerospace system telemetry. In: Deshmukh, J.V., Havelund, K., Perez, I. (eds.) NFM 2022. LNCS, vol. 13260, pp. 527–537. Springer, Cham (2022). https://doi.org/10.1007/978-3-031-06773-0_28

25. Montanari, A., Ratto, E., Corsetti, E., Morzenti, A.: Embedding time granularity in logical specifications of real-time systems. In: Proceedings of EUROMICRO 1991 Workshop on Real-Time Systems, pp. 88–97 (1991)

26. Mostafa, M., Bonakdarpour, B.: Decentralized runtime verification of LTL specifications in distributed systems. In: 2015 IEEE International Parallel and Distributed Processing Symposium, pp. 494–503 (2015)

27. Okubo, N.: Using R2U2 in JAXA program. Electronic correspondence, November–December 2020. Series of emails and zoom call from JAXA to PI with technical questions about embedding R2U2 into an autonomous satellite mission with a provable memory bound of 200 KB

28. Ouaknine, J., Worrell, J.: Some recent results in metric temporal logic. In: Cassez, F., Jard, C. (eds.) FORMATS 2008. LNCS, vol. 5215, pp. 1–13. Springer, Heidelberg (2008). https://doi.org/10.1007/978-3-540-85778-5_1

29. Reinbacher, T., Rozier, K.Y., Schumann, J.: Temporal-logic based runtime observer pairs for system health management of real-time systems. In: Ábrahám, E., Havelund, K. (eds.) TACAS 2014. LNCS, vol. 8413, pp. 357–372. Springer, Heidelberg (2014). https://doi.org/10.1007/978-3-642-54862-8_24

30. Rozier, K.Y., Schumann, J.: R2U2: tool overview. In: Proceedings of International Workshop on Competitions, Usability, Benchmarks, Evaluation, and Standardisation for Runtime Verification Tools (RV-CUBES), Seattle, WA, USA, vol. 3, pp. 138–156. Kalpa Publications, September 2017

31. Schumann, J., Moosbrugger, P., Rozier, K.Y.: Runtime analysis with R2U2: a tool exhibition report. In: Falcone, Y., Sánchez, C. (eds.) RV 2016. LNCS, vol. 10012, pp. 504–509. Springer, Cham (2016). https://doi.org/10.1007/978-3-319-46982-9_35

32. Wolper, P.: Temporal logic can be more expressive. Inf. Control **56**(1), 72–99 (1983)

Author Index

Printed in the United States
by Baker & Taylor Publisher Services